BRACKETING THE ENEMY

BRACKETING THE ENEMY

Forward Observers in World War II

JOHN R. WALKER

UNIVERSITY OF OKLAHOMA PRESS : NORMAN

Publication of this book is made possible through the generosity of Edith Kinney Gaylord.

Library of Congress Cataloging-in-Publication Data

Walker, John R., 1946-
 Bracketing the enemy : forward observers in World War II / John R. Walker.
  pages cm.
 Includes bibliographical references and index.
 ISBN 978-0-8061-4380-4 (hardcover : alk. paper)   1. World War, 1939–1945—Artillery operations, American.   2. World War, 1939–1945—Regimental histories—United States.   3. Artillery, Field and mountain—United States—History—20th century.   4. United States. Army. Infantry Division, 37th.   5. United States. Army. Infantry Division, 87th.   6. Unified operations (Military science)—History—20th century.   I. Title.
  D769.339.W35 2013
  940.54'1273—dc23
                                                              2012050123

During World War II, the U.S. Army created the Combat Infantryman's Badge to foster esprit de corps among infantrymen who served in battle. Forward-observer teams served in the thick of battle, but as artillerymen, they were ineligible to receive the award. Thus, the army never recognized forward observers as combatants. Recently, the army has created the Combat Action Badge, similar to the CIB, to signify that the wearer has served in combat. Since then, many forward observers have earned this new award.

While raising the morale of servicemen was important during World War II, perhaps the real significance of the forward observers lies not in the fact that they were active combatants, but that they provided the human element previously missing to provide close artillery support of infantry and in so doing achieved true combined-arms effectiveness. This study is dedicated to all U.S. artillerymen, officers, and enlisted men who performed forward-observation duties during World War II.

Contents

Illustrations

Chart and Maps

Preface and Acknowledgments

In August 1967 I was a twenty-year-old finance clerk serving with the U.S. Army's 1st Infantry Division at Di An, Vietnam. One day I found a plastic Combat Infantryman's Badge (CIB) someone had left on the counter where I worked. My father, Donald L. Walker of Alliance, Ohio, was a combat veteran of World War II. I wrote my Dad a letter and asked him if he knew what a CIB was. He replied that an infantry officer had awarded CIBs to his lieutenant and to him after they had breached the Siegfried Line in Germany in February 1945. The award was unofficial because only combat infantrymen were authorized to wear them, but it pleased him to know that at least one infantry officer acknowledged that field artillery forward observers were in the war too. My father passed away in 1990, but I never forgot the words in his letter.

The idea to conduct a detailed study of forward observers first came to me six years later, after having met my father's former lieutenant, James R. McGhee of Mount Vernon, Illinois, at an annual reunion of the 87th Division Association. He had served as a forward observer with Battery A, 334th Field Artillery Battalion in that division. Jim McGhee; Gayle A. Bricker, Jr., of Sarver, Pennsylvania; and my father were together for more than two months as a three-man forward observer party during the initial stages of the division's combat experience in late 1944 and early 1945.

On February 12, 1945, Gayle Bricker left the unit and was evacuated to a field hospital after suffering internal injuries from concussion.[1] After the 87th Division breached the Siegfried Line in February 1945, Lt. Col. Robert Moran, commanding officer of the 3rd Battalion, 345th Infantry Regiment, awarded CIBs to two field artillerymen, Jim McGhee and my father. These were unofficial, yet they represented one instance where an infantry officer recognized that artillery forward observers were active combatants, not simply support troops coming under fire. Five decades later McGhee and a few other veterans petitioned the Department of the Army to create a similar award for artillerymen. Retired army general William C. Westmoreland, who had begun his military career as an artillery-man in World War II, wished the project well but called this effort "unsalable."[2]

Although this drive failed, McGhee's unsuccessful efforts revealed how few people know what is meant by the term "forward observer" and that the history of these servicemen is virtually unknown. More importantly for scholarship, historians have thus far overlooked the vital role that forward-observation teams on the ground played in pro-viding close-fire support to maneuvering infantry during World War II. This is particularly significant because the technology enabling field guns to hurl shells long distances had existed for nearly eighty years before field artillery devised a system able to provide indirect-fire support in response to emerging battlefield situations.

Boyd L. Dastrup, U.S. field artillery command historian, describes forward observers as "the unsung heroes of the field artil-lery and generally expendable lieutenants," adding, "without them, the field artillery would have been useless."[3] While artillerymen on the ground did not provide all observation for field artillery fire direction during World War II, there were many times when their work was indispensable.

This is a study of one aspect of the combined-arms doctrine that evolved before and during World War II. Tactics involving the use of infantry with supporting artillery matured during that con-flict.[4] Its focus is on the theory and practice of keeping forward-observer teams with maneuvering infantry to find targets and adjust artillery fire.[5]

The historiography of World War II covers a wide range of top-ics that examine almost every military aspect imaginable. Historians have conducted much scholarly research on the subject of field artillery, including new developments in artillery weapons, ordnance, and tactics, such as the transition from direct to indirect artillery fire.[6] Except for a small number of published memoirs, however, the history of forward observers has been virtually overlooked because primary sources are scattered and hard to find.[7] The National Archives repository in College Park, Maryland, for example, holds volumes of official army records from the war. Yet in the majority of these documents, the specific actions of forward observers appear only sporadically and rarely identify participants or details. Dastrup indicates that even by examining battalion histories and after-action reports, little information about these men can be gleaned, adding: "There is good reason why little has been written about forward observers. The information is scarce."[8]

A notable exception to this is each division's set of general orders issued for medals and citations. These provide a brief but detailed narrative account of what a soldier did to receive his medal. Each citation is a history in itself. Some are more detailed than others, but a reader gains from them a fairly clear idea of what happened. Narratives appear most frequently for the Bronze Star for heroic achievement and the Silver Star for gallantry in action. Because the designation "forward observer" was not added to the army's Table of Organization and Equipment until 1944, most official records fail to identify or even acknowledge such men.[9] For example, unit and division histories may record details regarding forward observers without identifying individuals by name. Various personnel rosters for field artillery units fail to identify who served in this capacity. Without the names, it is difficult to know if two different sources are describing the same incident, or exactly what was involved and what a soldier was doing. Identifying the individuals involved makes it easier for the historian to corroborate the facts given from different sources and lends itself to accuracy. So, although official army records from World War II are plentiful, in general, the orders for medals provide the best source for identifying forward observers and recording their actions explicitly.

Probably the next-best primary source is the *Field Artillery Journal*. From 1911 until it became *FA Journal* and finally *The Artillery Journal*, this publication has been the professional journal of the U.S. field artillery. It provides a number of articles written by professional artillerymen describing the development of and technicalities involved in achieving successful artillery-infantry coordination in combat. In addition, those from the interwar period provide many important details of the new developments from that decade as well as the idea of positioning forward observers in the front lines, a new concept at that time. During World War II, numerous forward observers submitted articles for publication, describing their personal experiences in combat and indicating what worked and what did not in action.

Because professional soldiers have written more about the development of combined-arms tactics than others, the historical literature is found mostly in military repositories and libraries such as the excellent Combined Arms Research Library at Fort Leavenworth, Kansas, and the U.S. Army Military History Institute Library at Carlisle Barracks, Pennsylvania. Yet some published works may be found on the shelf in local public or university libraries. Former army officer Jonathan M. House may be regarded as the foremost authority on the history of combined arms in the United States. His study, written in 1984, for the U.S. Army Command and General Staff College at Fort Leavenworth and then later published as a book, is the foundation from which others have subsequently drawn.[10] British historian and former artillery officer J. B. A. Bailey has also published an important book on field artillery and firepower.[11] From the Revolutionary War to the first strategic-bombing campaigns of World War II, it would be difficult to imagine a single more-destructive weapon than artillery. Cannons had a longer range than small arms and did not require the same degree of accuracy to inflict harm on an enemy. A big gun firing a shell loaded with canister or grape shot had the same effect as a gigantic shotgun propelling a sizable cluster of heavy pieces of metal that fanned out as they moved forward. The spread of canister at three hundred feet was a circle thirty-two feet in diameter with enough impetus to kill or severely wound anyone in its path.[12] One shell fired into the midst of a bunched-up enemy did not have to be aimed at any individual soldier to take advantage of its deadly and destructive force. Although the American army's use of grape shot and canister dates back to the Revolutionary War, field artillerymen of the nineteenth century failed to use their combat arm to its fullest tactical advantage. World

War I was the first major war in which artillery caused the greater number of casualties.[13]

The present study, after an overview of the history of artillery starting in the mid-nineteenth century, the long transition from direct to indirect artillery fire, and the results of that transition in World War I, follows the experience of one U.S. infantry division from the Pacific and one from the European theater during World War II to describe the practice of combined-arms doctrine at the tactical level through the use of forward-observer teams accompanying maneuvering infantry. By studying a division in each of these two theaters, one can compare American combined-arms tactics against the Japanese and then against the German practice. Throughout this study, unless otherwise noted, the terms "forward observer" and "forward observers" will mean any member of a forward-observation party regardless of rank.

Chapter one begins with a quick overview of the history of artillery, particularly U.S. Army field artillery, and a description of American combined-arms doctrine and tactics during the late nineteenth century. The many major improvements in small-arms fire, coming almost simultaneously with the increase in the range of field guns, had a number of implications for the use of indirect fire. As Vardell Edwards Nesmith, Jr., has noted, at times the technology of field artillery often exceeded its efficient application.[14]

The new weaponry used during World War I forced all the major belligerents to use indirect artillery fire or suffer the loss of their firing batteries. No longer could they position their field guns where the enemy could see them. The howitzers and big guns of that day were extremely accurate. Shells could land on targets with extreme precision if the exact location of the target could be calculated in relation to the guns. Yet unless the artillery could train on targets as they appeared and when infantry had its most crucial need for fire support, their extreme precision and accuracy could not realize its full tactical effectiveness. To rectify this required someone who was always in position to observe the target, had the authority to command and control the direction of artillery fire, and had the means to communicate the results.

It was the Field Artillery School during the 1930s that made the most appreciable advances in the control of artillery fire that would serve the army well during the next war. These changes enabled artillery to provide the responsive, close, and effective fire support to infantry that was potentially available but consistently lacking during World War I.

Chapter two introduces the two divisions used for case studies in this work: the 37th "Buckeye" Division, Ohio National Guard, in the Pacific and the 87th "Golden Acorn" Division in the European theater. One of the first units to experiment with motorized artillery was the 135th Field Artillery Regiment, Ohio National Guard. Aside from this, the 37th Division makes a good subject for study in the Pacific because its commanding officer, Maj. Gen. Robert Beightler, was a firm believer in the use of artillery to try to minimize infantry casualties. The 37th Division and the 1st Cavalry Division shared the distinction of being the only two U.S. Army divisions to fight a major battle in a large urban setting in the Pacific.

The 87th Division arrived in Europe during the latter stages of the war, yet it saw its share of hard fighting in the five months that it was in combat. Brig. Gen. William W. Ford commanded the division's artillery. He had been instrumental in convincing the army of the value of using light single-engine airplanes for aerial observation and for directing artillery strikes.

The third chapter compares Japanese and German combined-arms tactics and how these measured up to those of the U.S. Army. It also reviews their military doctrine, tactics, and equipment and contrasts them with the Americans. It compares their field artillery to American field artillery and makes a similar analysis of secondary factors in each theater.

Because some might regard various comments appearing in this book regarding the Japanese as racially biased, before continuing to chapter four, a few points are offered here for consideration. Some writers argue that the war in the Pacific was a racial conflict and that American historians tend to view the performance of the Japanese armed forces through a racially biased lens. Certainly in many instances a mutual racial hatred existed between American and Japanese forces fighting each other. Whether most if any historians from these nations write about the Pacific War with any intentional bias is a little harder to discern.

The appellation most Americans used for the Japanese during World War II was "Jap" (a term that appears frequently in quotations in chapters four through six). One may argue that this was a racial slur, and perhaps it was intended as one then. Yet in the absence of any malice, one might regard it as a shortened version of "Japanese," or "Nips" for Nipponese, since Nippon is the true name of the Japanese nation in their own language. The object here is not to rekindle any feelings of ill will but simply to employ the same language in use at the time.

The many points made throughout this study are limited to a comparison of Japanese and American military effectiveness as defined by facts that include logistics, equipment, and tactical effectiveness. It is difficult to inject any racial bias into a comparison of weapons like the Japanese standard-issue bolt-action rifle and the semiautomatic M-1 rifle. Likewise, whether one views the Japanese infantrymen as fanatical or extremely brave does not matter in an analysis of tactical doctrine and combined-arms effectiveness. In countless Pacific campaigns, the vast majority of Japanese defenders died not because they had to but because it was in keeping with what they had been taught. That in itself is a cultural issue, one that had bad consequences for the Japanese. But if one doubts the valor of individual Japanese soldiers, the U.S. Army's 442nd Regimental Combat Team, comprised of Japanese Americans, stands as a stellar example of the outstanding fighting capabilities to be realized when Japanese bravery is combined with American military doctrine and technology. Fighting in Europe, the 442nd won more medals in the war than any other unit in U.S. history.[15]

Moving on, chapter four follows the experience of the 37th Division's artillery on New Georgia. This was the division's initial combat experience, relatively short lived in comparison to what would follow. Climate, topography, and terrain in many cases negated the advantages offered by the Americans' equipment and training, forcing them to use new tactics such as adjusting artillery fire by sound.

The fifth chapter covers the fight for Bougainville. This campaign saw the 37th Division take part in heavy fighting both on the offensive and defensive. It thoroughly tested the ability of the forward observers to maintain control and communications. In addition, artillery showed just how deadly the massing of fires could be, with the Americans employing interservice joint fires very effectively.

Chapter six reveals the dangers inherent in modern warfare when weapons of mass destruction are used in combat in proximity to a large civilian population. Urban fighting was new to the 37th Division, and something it had not trained to undertake. The Japanese had more artillery available on Luzon than they had on New Georgia or Bougainville but used it poorly for a variety of reasons. The battle for Luzon also revealed that by this late stage in the war, U.S. combined-arms effectiveness in the Pacific had improved somewhat between infantry and armor and infantry and tactical air support.

The seventh chapter looks at the first combat experience of the 87th Division against the German army. What becomes immediately apparent is the more pronounced similarity between these two forces at the tactical level than existed between combatants in the Pacific. The Golden Acorn Division was fighting an enemy with much more experience in the execution of combined-arms warfare than it possessed. For a "green" division, the Saar Valley experience was a difficult struggle against an intelligent and sophisticated foe.

Chapter eight covers artillery-infantry combined-arms effectiveness in the Ardennes Offensive, known commonly as the Battle of the Bulge. Here American artillery played a key role, first in preventing a German breakthrough to the coast, then in reducing the Bulge. Bad flying weather and heavily wooded terrain meant that the forward observers on the ground made a major contribution at very crucial times in turning the tide of the battle. Ground observers became so thoroughly integrated with the infantrymen they were protecting that they frequently took command of small-unit actions when the infantry's junior officers became casualties.

Chapter nine follows the 87th Division as it smashed through the West Wall, crossed the Rhine, and raced across central Germany to find itself near the Czech border when the war in Europe ended. Penetrating the heart of the Reich forced the Americans to utilize everything they had learned from their short but intense combat experience. Not only did this campaign blur the distinction between artillerymen and infantrymen on the front lines but also the distinction between officers and enlisted men. Enlisted men of all grades stepped in to do the job when their officers became casualties.

Chapter ten describes how four field-artillery forward observers earned the Medal of Honor in three wars. What some of them did to receive that award is some indication of how much infantry company commanders had come to rely upon their forward observers.

The actions of these four artillerymen also reveal how thoroughly integrated with infantrymen they became in combat.

The eleventh chapter presents the conclusions of this study, including an epilogue. Certainly, the artillery-infantry team was not the only combat-arms duo to achieve combined-arms effectiveness during World War II, but it was arguably the key to U.S. success in ground combat. Each artillery piece was, as historian Gerald Astor has written, "a tiny spark in a long night."[16] The purpose of this study is to reveal the important but largely unknown role forward observers played in guiding and directing that enormously powerful spark.

The author would like to acknowledge all who have helped in so many different ways. Thanks to Donald Singer and the staff in the Modern Military Records, Textual Archives Service Division of the National Archives and Records Administration in College Park, Maryland, for the help I received while there. My deepest gratitude to Kathy Buker, special-collections librarian, and Ed Burgess, director, of the Combined Arms Research Library at Fort Leavenworth, Kansas. This is a wonderful research facility with a staff dedicated to helping the public.

Thanks to Boyd L. Dastrup, command historian, U.S. Field Artillery, Fort Sill, Oklahoma, for his kind encouragement, suggestions, and counsel. I am also very grateful to Michelle Youngblood, former staff member of the Morris Swett Technical Library in the Field Artillery School at Fort Sill, who was particularly helpful during my visit. Thanks to the U.S. Field Artillery School for kindly permitting me to witness firing exercises and for allowing me to sit in on a class for forward observer trainees. Thanks also to Richard Baker and the staff at the U.S. Army Military History Institute at Carlisle, Pennsylvania, for their kind help during my research there.

I am grateful to the 37th Division Veterans' Association and the 87th Infantry Legacy Association. Both granted me permission to use unlimited material from their respective division histories. This information is indispensable for providing an historical overview and for accurately corroborating the details of individual actions that took place within these units with other primary sources.

My former fellow University of Akron history classmates and co-employees, Don Appleby with Bierce Library and John Ball with the University Archives, provided much-welcomed assistance. During the course of this endeavor, they have given me a maximum of help and support by obtaining books through Ohio Link and interlibrary loan. In addition, Don and John have given me valuable encouragement and advice. Former artilleryman and Vietnam veteran Richard Diser, of Alliance, Ohio, critically reviewed the earliest versions of my chapter drafts.

Herzlichen Dank to my good friend, Guenter Nieser, and to his good wife, Lucia. My wife, Alice, and I met them during a trip to Germany in May 2005. They kindly took us in their auto from their home in Fechingen to Medelsheim and the Schnee Eifel and patiently guided us over the former battleground. Thanks also to our other German friends, Wolfgang and Monika Neutsch, for their help.

I must also acknowledge my daughter, Alice Jane, who patiently explained to me how to reformat my files to make them suitable for publication. Special thanks to my dear wife, Alice, for her encouragement, common-sense advice, patience, and forbearance.

The most important contributors to this book are the many veterans of World War II who shared with me their memories and the historical details of their personal experiences. My contact with each and every one of them was a privilege and honor.

BRACKETING THE ENEMY

Introduction

To understand the need for a forward observer, one must understand that by the time of the Civil War, artillery technology had begun to outdistance doctrine and tactics, which had served the army well to that point. Historian Robert Remini, for example, has debunked an old myth, insisting that it was American artillery, not sharpshooters, that won the Battle of New Orleans for Andrew Jackson in January 1815, slaughtering British veterans of the Napoleonic Wars by the hundreds on the Chalmette battlefield.[1] And military historian John S. D. Eisenhower considers nineteenth-century field artillery to be the U.S. Army's best-trained and most technologically advanced combat arm of the antebellum era. Writing about the Mexican War of 1846–48, he observes that American troops repeatedly went into battle overwhelmingly outnumbered by the Mexicans, concluding that the superiority of U.S. field artillery repeatedly defeated the enemy and in essence won that war.[2]

By the Civil War, cannons could easily hurl a shell more than two miles, yet artillerymen continued to take aim at their targets as they had for centuries.[3] Although shots fired out of sight might strike the enemy, they could also hit friendly troops or anyone else who strayed into the target area. Hence the advantages accruing to longer range were wasted while the technical capability of being able to shoot at something unseen created several unanticipated problems. In time, this led from the old system based on direct fire, where the gunners could see their targets, to one of indirect fire, where they could not. Bailey describes indirect fire as "the most important conceptual and technical innovation in over five hundred years of artillery practice."[4] Yet during the Spanish-American War, U.S. field artillery still used direct fire.[5]

Not until World War I did armies first come to rely on indirect artillery fire. The longer range, increased accuracy, and higher velocity of small arms—that is, rifles and automatic weapons—meant that even from a distance, advancing infantrymen could easily spot and kill the gun crews of unconcealed batteries, thus forcing artillerymen to use indirect fire.[6]

During that conflict, some artillerymen referred to as forward observers were stationed among infantrymen at or near the front lines. But as the infantrymen advanced, these men stayed behind and thereby lost the ability to provide continuous supporting fire in response to new targets as they became known. When unable to see the target, battery commanders relied upon methods of unobserved fire.[7] In fact, the majority of artillery fire during World War I was unobserved.

Early attempts by the major armies to routinely use indirect fire in combat often had unintended if not disastrous results. French general Alexandre Percin estimated that his army's artillery killed 75,000 of its own soldiers during the war.[8] Although one may question the accuracy of this estimate, the number of deaths resulting from friendly fire, or "amicicide" as author Charles R. Shrader calls it, was recognizably a significant problem in World War I.[9] A better method of combining the combat effectiveness of infantry and artillery was needed.

Trying to learn from these mistakes, the U.S. Army convened a number of postwar boards to examine the performance of its field artillery in the war and to examine ways to improve combined-arms effectiveness between all combat forces. Key developments for field artillery during this time included motorization, creation of the fire-direction center, the advent of practical field radios, and taking responsibility for observation and conduct of fire away from the battery commander and giving it to an artillery observer, either on the ground or in the air.[10] Although members of gun crews, fire-direction centers, and other artillerymen are all human, the forward observer represented the only human element in the system who stays at the scene of the action, or otherwise continually has in sight the target under fire. This was an entirely new configuration in the practice of indirect fire.

Prior to its use in actual combat, no one was sure how well the new methods would work, or if they would even work at all. Several of those surveyed by the Hero Board indicated that if forward observers were positioned too close to combat, they would either be killed or unable to function. But the system that was developed worked so well that infantrymen eventually came to regard their supporting artillery as a "utility." A quick call by a forward observer over a field radio or telephone summoned a barrage of artillery shells that could quickly eliminate an enemy outpost and allow the soldiers to proceed without harm, much like turning on the water hose to put out a fire.[11]

A forward-observer party consisted of one commissioned officer (the designated forward observer) and at least two enlisted artillerymen. The field radios of the time were so heavy that it took two men to carry each half to an action post. After reassembly, one man served as radio operator while the other helped observe.[12] When they sighted a target, the radioman called the fire-direction center for a fire mission, and the forward observer gave the instructions. After the first round had landed, left, right, in front of, or beyond the objective, standard procedure was to adjust the second round so that it would fall opposite from the first one and equidistant to the target. Thus having "bracketed" it, the third round was intended to split the difference and hit the mark. A good forward observer could adjust fire with amazing accuracy.

During World War II, field-artillery pilots flew light, single-engine, unarmed planes, with an artillery officer on board who might also serve as a forward observer on the ground. Their mission was to find targets and direct fire missions. Since observation from a moving airplane gives one a much wider perspective and a better view of potential targets, the number of field-artillery strikes directed from the air far exceeded the number on the ground. Although this was true, there were more observers on foot than in planes. It was common practice for a direct-support field-artillery battalion to furnish one forward-observer party for each supported rifle company up to a total of nine. In contrast, an entire field-artillery battalion was allotted two observation planes, two pilots, and two observers; in the

Pacific theater, pilots flying alone generally directed fire missions themselves.[13]

The use of aerial observers, however, did not eliminate the constant need to keep men on the ground with the forward elements of the infantry. Bad weather often grounded planes, and poor visibility as well as various topographical features put aerial observers at a disadvantage for spotting enemy activity. Inclement weather also occurred at some of the most critical times of the war, most notably during the early phases of the German Ardennes Offensive of December 1944.[14]

Because forward observers were with the riflemen on the ground, there was no interruption in their ability to observe targets if weather conditions kept observation planes from flying. At least they could see as much as the men they accompanied, if not more. Aerial observation did have many advantages, and pilots were able to help the infantrymen in other ways besides providing fire support, such as delivering ammunition and supplies or by interacting with friendly forces in a number of ways.[15] But artillery observers on the ground were such an integral part of the units they were with that they often became nearly indispensable to company commanders.[16]

In the early phases of World War II, only commissioned officers acted as forward observers. But as the fighting continued, so many of the officers became casualties that it was not unusual for the enlisted men accompanying them to take on that role.[17] Although many of these men became casualties, an accurate assessment of how many artillerymen performing forward-observation duties became casualties during the war would be difficult to compute.[18] There is no official army record that consistently identifies forward-observation personnel from 1941 to 1945. Sometimes the duty rotated. More often, it changed only as various members of such parties became casualties.

Forward-observation duty was dangerous work. It was particularly hazardous because these men were normally close enough to the target to be seen by the enemy. Also, if they misjudged their position, they could inadvertently bring friendly fire down on the very men they were trying to protect, or themselves.[19] Thus, it was a very hazardous job laden with tremendous responsibility, so much so, in fact, that by 1944 the army declared that in ground combat "the artillery forward observer is potentially the most powerful individual in the forward area."[20]

Infantrymen appreciated and have frequently acknowledged the forward observer's capabilities as a force multiplier, though not expressing it quite that way. Ralph Carver of Erie, Pennsylvania, was a platoon sergeant with Company D, 345th Infantry in Europe. After the war Carver noted: "It did not take the combat infantry troops long to realize the importance of coordination with artillery. Most of us, at one time or another, probably owed our lives to artillery support. They helped prepare the way during the attack, broke up or suppressed enemy counter attacks, and helped cover our defensive positions. They also shared our wet, muddy, cold foxholes, our misery, our fears, and the courage it often took to just keep going. The FO teams along with the medics, soon had the respect and highest regard of the infantry."[21] Retired Marine Corps general Bernard E. Trainor expressed much the same sentiment in even-more-personal terms when he wrote: "Having had combat commands in two wars I have high regard for FOs and owe my life to them on more than one occasion."[22]

Some may regard reference to such comments as singling out the forward observer for undue praise. In terms of conventional war, no single branch of service, combat arm, or weapon wins a war by itself. Historian Gerald Astor made an excellent point for all students of military history to bear in mind when he observed that war is "a group enterprise" involving a countless number of acts by a multitude of people, including infantrymen and artillerymen.[23] When Gen. George S. Patton declared that artillery had won World War II, what he was really alluding to was its power as a force multiplier. Artillery observers both in the air and on the ground provided the human element necessary to help achieve that success. Those on the ground filled in the gap left when their aerial counterparts either were unable to fly or could not see targets. They also became integrated with fighting forces on the ground in ways that those flying above could not.

CHAPTER 1

The Evolution of Forward Observation

For the first nine hundred years of the history of artillery, direct fire was the main way of using field guns in battle. The term "direct fire" means that the men aiming the weapon could observe their intended target, in contrast to "indirect fire," where the intended mark could not be seen. Certainly over this period of time, there were numerous instances when armies fired projectiles over the tops of trees, hills, and walls at an enemy unseen. The effectiveness of such shots generally was greater in those circumstances where an enemy remained confined within some limited space such as a walled fortress.

During the era of direct fire and even to this day, the ability to mass bombardments effectively has depended to a large extent on what Jonathan House calls the three "Cs" of combat: command, control, and communications. Historically, mobility has also been a key factor. Napoleon's use of horse or flying artillery was a prime example of this. Through their rapid mobility, he could mass fire quickly to generate overwhelming destruction at a decisive point to support his main effort.[1] Napoleon typically placed his artillery a few hundred yards behind the infantry it was supporting, sometimes even ahead of it. From this advance position, his gunners could see their targets and exert a psychological effect on both his army and the enemy's.[2]

Nineteenth-century field guns included the cannon, the howitzer, and the mortar. The cannon has a flat trajectory, while that of a howitzer follows an arc. The trajectory of a mortar is a very high-angle parabola. The mortar and the howitzer enable an army to rain down shells on an enemy positioned in or behind concealment, while the trajectory of a cannon shot cannot do that. Because most early artillery fire involved aiming the guns at the targets, the armies of the world relied upon the cannon.

By the time of the American Civil War, the effective range of cannons had reached a distance of over a mile, making it frequently impossible for the gunners to see their targets unless they were sitting on top of a high hill or if their enemy appeared before them across a long, flat, open stretch of terrain.[3] Yet there is no advantage to longer range if an artilleryman does not know if he has hit his intended target. To use indirect fire effectively required an observation and communications system that would allow for aiming adjustments. In short, it required an observer forward of the guns.

The introduction of the rifled musket around this same time was an early contributor to the eventual transition to indirect artillery fire.[4] Tactical offensive doctrine of that day called for cannons to be positioned in front of lines of infantry, aimed for a range of about 300 yards, and fired directly into ranks of enemy infantry. While smoothbore projectiles from small arms could carry farther than 50 yards, the accurate range of opposing infantry was not much more than that, leaving cannoneers relatively safe from enemy musketry.[5] The introduction of rifles with increased range and accuracy put cannoneers near the front lines at risk from infantry fire. How much this altered the tactics used in that war is another question, however.

The prevailing orthodoxy among historians is that rifled muskets with a range accurate up to 1,000 yards tipped the balance of power between attack and defense on the Civil War battlefield. How many first-time enlistees might have been able to hit a human target beyond 250 yards consistently, let alone 1,000 yards, is questionable. Recent research, however, suggests that the practical effect of rifled muskets on combat was far less than it was previously thought to be. In the days before motorized field artillery, horses supplied the power to move field pieces. The availability of horses, then, was a prime factor in determining the mobility of artillery, and because the Confederate army had fewer horses than the Union army, its artillery was less effective.[6] During World War II, German and

Japanese artillery would suffer the same lack of mobility for much the same reason. In any case, rifled muskets unquestionably increased the effective range of infantry small-arms fire. They and the high-powered bolt-action rifles that would soon follow forced armies over time to consider placing their batteries beyond enemy small-arms range or risk losing their gun crews or horses during battles.

The increased range and accuracy of the rifled musket also hastened the tendency to dig trenches, although not until the latter stages of the Civil War did this become more common. Generals from both sides came to appreciate the difficulty of making frontal assaults against fortified troops and encouraged their own armies to dig trenches.[7] As this practice became more widespread, the overall effectiveness of direct or low-angle flat-trajectory artillery fire decreased since cannoneers were unable to see a dug-in enemy. So the appearance of trenches for protection and concealment also contributed to the eventual switch from cannons to howitzers and from direct to indirect fire.[8]

The history of forward observers in the United States also began during the Civil War. Probably the first recorded instance of using a forward observer to direct artillery fire occurred in May 1862. Maj. Albert J. Myers used signal flags from a tugboat to direct Union artillery against a coastal fort held by the Confederates at Sewell's Point near Norfolk, Virginia.[9] Signal Corps balloons were used in similar fashion throughout the war.

J. B. A. Bailey claims that both North and South started the war with no coherent doctrine, adding that American artillery saw its own importance significantly reduced by "a mismatch of received tactics and new infantry technology." He does note that although field artillery was not the predominant arm during the Civil War, at certain critical times it played a clearly decisive role in the outcome of particular engagements.[10]

The Battle of Malvern Hill on July 1, 1862, provides a good example clearly illustrating the advantages of massed fires used defensively. Confederate generals Robert E. Lee and James Longstreet attacked Union positions over a 1,500-yard front without consulting their chief of artillery, Brig. Gen. William Pendleton. Lee and Longstreet had presumed they would have the support of 120 guns, but only 16 participated in the artillery duel preceding the assault; these were answered by the fire of 50 Union cannons.[11] The North's artillery reserve played an important role in the battle, decimating the Confederate offensive. Maj. Henry Hunt, Union chief of artillery, used every available gun at his disposal. Twenty-five Union batteries caused more than half of the 5,000 Confederate casualties. By the efficient use of his resources, Hunt helped destroy the attackers at Malvern Hill and established the artillery reserve as an important artillery organization.[12]

Although defensive artillery played a positive role for the North at Malvern Hill, it had its negative side as well: friendly fire killed many Union soldiers that day. Army of the Potomac cannoneers fired over their own forward lines even though the soldiers were close to the enemy. This was considered to be a safe practice assuming that no ammunition was defective, every battery knew the precise location of its targets, the gun crews had set their weapons at the proper elevation, and the men never cut a fuse too short. If even one assumption proved false, shots could fall short. As the rate of firing increased, heavy smoke enveloped the battlefield. Consequently, Union fire most likely fell among friendly troops. How many is not as important as the fact that it happened. To make matters worse, Union naval gunfire also hit Federal troops. Around midafternoon, gunners onboard the USS *Galena* and *Mahaska* began firing at an enemy they could not see from more than two miles way. Shots from one or both ships landed close by Col. Robert O. Tyler's battery of the 1st Connecticut Heavy Artillery, killing or wounding a few of his men. Members of the Signal Corps promptly sent the message by way of flags: "For God's sake; stop firing!" This may be the only instance during the entire Civil War where officers of two separate services independently reported the same friendly fire incident.[13]

While the army's friendly fire incidents at Malvern Hill could be categorized as aimed fire used blindly, it nevertheless was a good example of the tragedy that could result when firing long-range artillery unobserved. It underscored how using indirect fire effectively would require a designated observer forward of the guns who could see the target and communicate the results of the fall of shots. The importance of massing fires would continue into the twentieth century, but the eventual transition to indirect fire would make it much more difficult to achieve successfully.

During the last thirty-five years of the nineteenth century, the U.S. Army divided its efforts between maintaining a frontier constabulary and numerous coastal garrisons.[14] Maintaining the coast artillery took precedence over field artillery, although the army did reestablish an artillery school at Fort Monroe, Virginia, in 1868. Important technological advances during these decades included wrapped-steel gun tubes, breech-loading guns, and modern carriages that absorbed recoil so that guns did not have to be repositioned after every shot, thus increasing the rate of fire.[15] Yet despite these various improvements and even greater increases in range, the field artillery, like that of other armies throughout the world, was reluctant to abandon direct fire.[16]

Throughout the Spanish-American War in 1898, U.S. field artillery continued using direct fire. During the campaign to capture Santiago, Cuba, Capt. Charles D. Parkhurst's gunners had difficulty hitting their targets because they could neither see the enemy nor tell where their shells were landing. Once, just as Parkhurst's battery prepared to fire shot at a position on the crest of a hill, he realized that American troops already had occupied the spot. In another instance, U.S. artillery could not break up an enemy counterattack because the guns rested on the reverse side of a slope—the gunners could neither see the enemy nor fire low enough to hit them because all their batteries consisted of high-velocity, flat-trajectory field pieces.[17] This episode demonstrated not only the suitability of howitzers for modern warfare but also the need for an observer to adjust indirect fire.

As the new century began, the major armies of the world remained committed to direct fire because of expediency and tradition. It was easier to use the old method than the new one.[18] But two wars that would erupt during the first decade of the new century would challenge the wisdom of maintaining that tradition. The Boer War and the Russo-Japanese War served notice to the armies of the world that they could no longer position their artillery in full view of the enemy.[19]

The Japanese use of artillery particularly impressed an American army officer who observed the Russo-Japanese War. Maj. Joseph E. Kuhn noted the emphasis the Japanese placed upon fire control. One officer directing all the siege artillery had telephone lines to observers who reported the results of each shot. They also used large-scale maps marked in one-centimeter grids, enabling them to fire by reference to specific grid coordinates.[20] In addition, they made an effort to conceal their batteries and to fire indirectly from behind cover, making it difficult for the Russians to detect the location of their guns.[21]

Japanese field artillery greatly influenced the thinking of the U.S. chief of field artillery, Brig. Gen. John P. Story, who attributed its success to their use of indirect fire. At the same time, American artillery underwent other major changes in organization and training. In January 1907 Pres. Theodore Roosevelt signed a bill separating field from coastal artillery.[22] In May the War Department reorganized the field artillery into six regiments, with two battalions per regiment, and incorporated it into the infantry division.[23] In 1910 the War Department assigned Capt. Dan D. Moore to Fort Sill, Oklahoma, to organize the School of Fire for Field Artillery.[24] In September 1911 the school began teaching its first class of artillerymen.

That same year Maj. William Snow, who would later become the first chief of field artillery, translated an article for *Field Artillery Journal* describing the system in which the forward observers communicated the results of the shot to the battery by using a series of arm motions and hand signals. This system could only work, however, if the artillerymen adjusting fire and the gun crews remained in sight of each other. For example, the observers indicated the shot was over if both extended one arm toward the target, or if one extended both arms and the other pointed one arm at the target.[25] Within a year, however, each field battery had three field telephones and each regiment and battalion two, which greatly improved the ability to transmit fire commands from the forward observer. But the one great drawback of telephones was that enemy shellfire frequently cut the lines.[26]

Not long after World War I began, static entrenchments replaced mobile operations due to the new intensity of infantry firepower, mainly from machine guns. So for the remainder of the war, field artillery became the most important weapon used to trump the effects of infantry firepower.[27] In defense, it could break up attacks, but if deployed too far forward the guns would be lost. This forced the transition from direct to indirect artillery fire in most combat situations.

In the early offensives, commanders expected massive artillery barrages to destroy enemy emplacements. These bombardments had a devastating effect on troops caught in the open. But attacks lost momentum when troops climbed out of the trenches and immediately lost contact with their supporting artillery. Maneuvering infantry had no way to shift or concentrate artillery support spontaneously on unanticipated areas of resistance or even to coordinate the timing of a preplanned barrage to keep pace with the momentum of an attack.[28]

The transition to trench warfare lead to what Bruce Gudmundsson calls "the great divorce" between infantry and artillery that took place between 1914 and 1915. Instead of looking to cooperate with each other, the two combat arms each looked to solve their own unique tactical problems.[29] This same "divorce" would take place in the American army after the United States entered the war.

By 1917 U.S. field artillery had realized major improvements in weaponry and gunnery techniques but was still mired tactically in its old role of making light cannons available to support infantry at close range. Rather than achieving a combined-arms effect that could be attained with artillery firing continuously in support of maneuvering infantry, the fire taught to be used was still direct. Infantry and artillery trained to fight separate battles.[30]

U.S. artillery doctrine vacillated between preparing to fight trench warfare or campaigns of maneuver. The American Expeditionary Force (AEF) commander, Gen. John Pershing, believed that trench warfare would eventually give way to what he called "open warfare" and that his troops must be trained to fight under both conditions. In the general's mind, the infantryman on the offensive with his rifle and bayonet was the sole key to victory, a view his staff accepted.[31] By 1915, however, artillery and machine guns had already demonstrated their tremendous killing ability on the defensive. When the United States entered the war, its military leaders faced a major problem of how to restore tactical mobility to the battlefield without losing men in the same numbers as the Europeans.[32]

Nevertheless, in 1917 the Americans still favored mobile warfare, which meant that they had to be able to shift artillery around the battlefield quickly using observed fire. The battery was the standard firing unit, and the battery commander directed and adjusted its fire. Typically positioned midway between the guns and the target, he ordered adjustments either by voice commands, telephone, or even hand signals. In addition, other artillerymen called "forward observation officers" stationed themselves in the infantry trenches or at designated observation posts.[33] Being closer to the target than the battery commander, these men conveyed the results of shots to him to aid his instructions. They also used the same methods of communications by which he contacted the battery.

The smoke and haze of battle obscured both the commander's view of his observers' signals and, in turn, the forward observers' ability to see the target, placing a much greater reliance on using vulnerable field telephones. German gunners soon realized this and tried to break the wires whenever possible. If the telephone lines were cut and the break could not be found, as often happened, then the batteries would depend on runners.[34] During a battle on July 18, 1918, the infantry and artillery units of the 26th Infantry "Yankee" Division had communication problems. As a result, they made heavy use of runners, who thus "died in droves."[35]

Regardless of what system the forward observer used, he was, in effect, tethered to his position either by telephone lines or his battery commander's line of sight. Therefore, when infantrymen did advance, the artillerists did not accompany them—the ability to provide continuous fire support at critical times was lost.

Because the United States fought World War I in concert with the Entente nations in the West, political circumstances compelled many American units to serve under French command and adapt to European methods of fighting. Yet even before this, French doctrine had heavily permeated American military thinking, and by 1917 this influence on field artillery was strong. French instructors taught at U.S. schools using French training manuals translated into English.[36]

It was rather natural then that the AEF would adopt the French liaison system. Prewar American doctrine had stressed the importance of using observed fire, but by the time the AEF entered into combat, Great Britain and France had already been fighting for nearly four years and were well accustomed to using map firing, rolling barrages, and gas-filled artillery shells as a means for compensating for the inability to maintain observed fire. The Western Allies had come to depend upon a system of preplanned and prescheduled fires.[37] The main function of the artillery liaison team was planning instead of adjusting fire once an action had begun.[38] What the French and British had discovered (and the Americans would soon learn) was

37th Division Howitzers, Camp Shelby, Miss., 1941. Photo courtesy of Donald L. Walker.

that once the infantry has left its trenches, its communication with artillery became tenuous, particularly if the telephones lines became severed. This, in turn, impaired the battery commander's ability to provide "effective fire on suddenly discovered objectives."[39] The ability of prescheduled bombardments to provide close support in combat was typically limited to the early stages of an action because, as many U.S. infantry officers noted, a "rigid schedule of fires is seldom the most efficient because an attack just does not progress as planned."[40] As House notes, "Massed fires were normally the result of carefully planned artillery concentrations, in which known targets were pre-designated on maps or overlays," all drawn up in advance.[41] But as one American officer explained, his supporting artillery was sufficient in situations in which details could be foreseen and planned in advance, otherwise "it fell down."[42]

The liaison-officer system called for each artillery regiment to send an officer to the infantry commander and each battery a sergeant to the regimental infantry battalion commanders.[43] Many battery commanders were reluctant to assign their best lieutenants to liaison duty and instead sent their least-experienced officers and those thought to be most expendable, doing little to instill infantry leaders with confidence in these men.[44] Similarly, enlisted artillerymen who created disciplinary problems could find themselves on the front lines. A half century after World War I, Robert Hiner of La Porte, Indiana, served with the 3rd Marine Division as a member of the crew on a 105-mm howitzer in Vietnam. Hiner recalled the case of one young artilleryman who got in trouble after going into a village and shooting some Vietnamese livestock. His commanding officer gave the offender the choice of standing before a court-martial or serving with a forward-observer team.[45]

The main shortcoming of the liaison system was its inability to provide responsive fire because the batteries either received no information from the front lines or got it in an untimely manner. Also, frontline infantry officers rarely notified anyone of their intended objectives because they were too occupied with the details of the fighting and were generally incapable of reporting their own positions precisely. Finally, artillery-infantry communications broke down either because of an inadequate supply of telephone wire or because the infantry had moved to an unknown location.[46]

Lack of observation due to an inconsistent communications or less-than-optimal positioning of observers resulted in poorly coordinated artillery support. Because the battery commander stayed at a post somewhere between his guns and the forward observers, his ability to communicate with one was enhanced at the expense of the other, thus weakening his ability to control fires in a timely and responsive manner. But this was only one facet of the problems of command, control, and communications that plagued commanders in action.

Higher command levels controlled artillery planning and deployment. Battery commanders and forward spotters who actually observed targets and could thus judge their importance were not involved in this process. Inadequate contact with the front lines meant field commanders were frequently unable to remain abreast of developing situations.[47] Thus, centralized artillery command coupled with poor communications resulted in much ineffective fire.

No one will ever know how much money was wasted on mistargeted artillery shells during World War I. Unless they caught troops in the open, barrages typically inflicted no serious damage. Artillery veteran Jay M. Lee described how most enemy harassing fire directed at American batteries that were spaced some distance apart fell harmlessly in between. Lee admitted that a few shells landed directly on protected dugouts and that there were several narrow escapes, yet "that compared with the tons upon tons of shells which came over and hit nothing but the ground, the actual casualties were relatively small."[48]

One infantry officer, Lt. George Hays, defined "observation fire" as "fire executed on targets picked up by artillery observation parties," noting that "during the World War, observation fire was practically negligible."[49] The routine use of unobserved fire had worse consequences than simply wasting ammunition as the number of friendly fire casualties rose. Artillery "amicicide" happened so frequently and routinely during the war that staff planners typically included some modest allowance for friendly fire casualties in their estimates of the human cost of an operation. One German field artillery regiment, the 49th, became known by German infantrymen as the "48 1/2" because of its reputation for firing short.[50]

Prewar U.S. doctrine stressed the need to use observed fire, but the static nature of trench warfare made it expedient for all the belligerents to rely on unobserved fire, at least until the front lines had advanced. Until the latter stages of the war, the lines rarely altered. After completing a survey and registering the guns, each side knew where the other lay in relation to friendly troops. So unless an attack resulted in a breakthrough or some change in position, there was not always a demand for observed fire. This led to the use of several methods of unobserved fire. One was called map firing and depended upon marking the targets on a map grid.[51] The British introduced the "rolling barrage," or what the Germans referred to picturesquely as "waltzing fire," in 1916. Under this system, the artillery laid down a curtain of shells that preceded the advancing infantry at a predetermined rate.[52]

Another artillery weapon that compensated for the lack of observation was poisonous gas. In 1916 the Germans developed specialized shells as a means of delivering poisonous gas to a targeted area without depending upon the prevailing winds. These shells did not need to hit a specific spot to be effective: they merely had to blanket the targeted area. Gas proved particularly effective in heavily wooded areas, for trees afforded no protection to troops from the harmful effects of gas while acting to slow the dispersal of toxic fumes.[53]

Military historian Mark E. Grotelueschen notes that using unobserved fire limited artillery's effectiveness to applications involving so-called set-piece operations, which generally obviated the need for close combined-arms cooperation.[54] In most set-piece attacks, the coordination of artillery was satisfactory, but in more fluid combat, it broke down. Close and continual observation of artillery fire and a reliable system of communication would have enabled the AEF to routinely exploit its gains, but these features were lacking. Using the 1st Infantry Division as an example, for nearly every attack made, if not all, the division used rolling barrages to achieve infantry-artillery coordination. Yet the barrages typically failed to overcome German resistance, requiring the infantry to contact the artillery for additional fire support. In the end, battalion commanders reported that "the infantry and artillery were unable to cooperate due to lack of liaison."[55]

Obviously, a reliable system of communications was lacking. Field telephones, though effective, were vulnerable to having their lines cut by enemy artillery bombardment or patrols. Radios were still heavy and bulky. No one had yet designed a portable set light enough for one man to carry, although many foresaw advantages to having one.[56] In 1916 a visionary American student of field artillery "suggested that a portable radio described in *Popular Mechanics* might be ideal for a forward observer 'if it could be made practical."[57]

Field artillery was a deadly combat arm during World War I, but because of its systems of command, control, and communications, it achieved its best results in the most static situations, while in those of mobile warfare, it was largely ineffective. Summing up the AEF conduct of the war, Grotelueschen writes: "Massive use of coordinated firepower, more than anything else, was the answer to the challenge of successful offensive on the battlefield."[58] After the war Lt. Col. Paul B. Malone, a veteran combat infantryman of the AEF, emphasized that the ability to provide almost instantaneous fire support to troops on the battlefield was the key component of any successful coordination between infantry and artillery. Additionally, Malone claimed, "The war proved the necessity of 'such cooperation' between infantry and field artillery that the two merge together 'into a single fighting unit.'"[59] Commanders did not realize then that even placing forward observers in observation posts at the foremost trenches did not put them close enough to targets to be able to adjust fire effectively; they needed to accompany the riflemen. But the reasoning was that, even if telephone lines remained open, placing the observer too far forward might result in his loss and thus the battery would lose all immediate observed-fire capability. British artillery regulations advised observers to stay far enough forward to be able to observe targets continuously during an attack but warned that it was "not usually sound for them to remain in the foremost firing line."[60]

After the war the U.S. Army convened several investigative boards to discuss the performance of American artillery in combat and to determine the changes necessary to increase its future effectiveness. These included the Hero, Lassiter, Westervelt (or Caliber), and Superior Boards.[61] Only the Hero and Superiors Boards really emphasized improving observation of fire or coordinating the actions of artillery with infantry.

Named for Brig. Gen. Andrew Hero, a board member, the Hero Board had the responsibility of studying AEF artillery combat operations in World War I. Most of the recommendations in its 840-page report addressed training and the optimal size of field guns, omitting any reference to improving artillery's ability to provide more-responsive fire support. The board expressed concern about aerial observation but for the most part ignored the problems associated with the widespread use of unobserved fire.[62]

The board did not entirely ignore the problem, however. One criticism involved the placement of forward observers, indicating that they were typically too far back to direct fire effectively. Brig. Gen. U. G. McAlexander addressed this issue, remonstrating how setting up observation posts wholly apart from the infantry and typically much further to the rear diminished effective coordination of effort between the batteries and those adjusting fire.[63]

With regard to ground observation, Schofield Andrews, G-3, 90th Infantry Division, offered three suggestions: first, competent artillery personnel should direct fire; second, good communications must exist between the observer and the batteries; and third, the observer at the front must have immediate authority to obtain the fire requested while eliminating all intermediaries. Andrews observed that in fluid combat situations, artillery lacked "a combination of observation and direction of fire by a trained artillery personnel [*sic*] and the authority to give fire at the front instead of at the rear." He also suggested removing fire control from the level of division and brigade command and relegating it to the battalion.[64] Although the official Hero Board report apparently overlooked Andrews's advice, by the time U.S. forces entered combat in World War II, battalions did control fire.[65]

During World War I, artillery liaison officers and their enlisted men had performed some of the duties that forward observers would handle during World War II. As they did, they sometimes found themselves in the middle of combat, and a significant number became casualties.[66] On July 18, 1918, a liaison officer of the 7th Field Artillery was with a detachment of the 28th Infantry as it assaulted enemy positions. After becoming separated from the group, he came under fire from a sniper. After trading shots with the German, he suffered a wound in the arm, thus ending his ability to provide liaison support.[67]

In the interwar years, some artillery officers thought that the experiences of these liaisons predicted what might happen to forward observers if they were positioned too close to the front lines. This may have been why Col. R. S. Abernethy suggested that artillery observers should work from a place that would subject them to "the least possible amount of unnecessary noise and confusion in the vicinity." In his opinion, the front lines represented the least advantageous positions for artillery observers, and he argued that they "should avoid posts that draw fire."[68]

In its final report, the Hero Board paid little attention to the problem of unobserved fire, instead recommending materiel solutions.[69] Neither the Lassiter Board nor the Westervelt (Caliber) Board gave any serious thought to improving observation methods or coordination of infantry with artillery.[70] Only the Superior Board addressed the subject of combined-arms tactics, completing its study on July 1, 1919. In the final analysis, it described close cooperation between arms so essential to success on the battlefield to be one of the primary lessons of the war. The board also recommended making artillery an organic part of each division.[71]

The real impetus for change in forward-observation techniques would come not from the top of U.S. Army command but from its lower echelons. From studying the lessons of World War I and applying what they learned to field exercises, the instructors and students of the Field Artillery School at Fort Sill would find a way to create a truly responsive system of fire support. Many older soldiers with long years of service would stubbornly resist these changes. But the thinking of two field-artillery officers, Maj. Carlos Brewer and Maj. Orlando Ward, would have a major influence on changes to tactical doctrine during the interwar years.

After a short period of transition in 1919 that saw four different Field Artillery School commandants in nine months, the appointment of Brig. Gen. Ernest Hinds to that post in October ushered in a new era of stability. The school then consisted of four departments: gunnery, tactics, materiel, and equitation. One of Hinds's first decisions was to shorten the length of time spent teaching technical subjects and increase the time spent teaching liaison and tactics.[72]

In 1922 the War Department combined a number of field-artillery basic officers' schools at Fort Sill into one and also transferred the school at Fort Bragg, North Carolina, to Oklahoma. By this time the emphasis on tactics was even stronger, with more than three-fourths of the total time allotted used for its teaching and firing exercises. The student curriculum emphasized liaison, communications, and fire direction.[73] Throughout the 1920s, training continued to emphasize tactics, though mostly the old system relying on the battery commander to adjust fire. By the latter half of the decade, however, the school began to consider new means of fire direction.

In the mid-1920s the Tactical Department began publishing *Field Artillery Notes* throughout the school year to provide updated guidance and instruction for artillery tactics and techniques. *Field Artillery Notes* began suggesting that someone other than the battery commander should direct and adjust artillery fire. Yet it overlooked the fact that he had the responsibility for computing the gunnery data for each of the three guns under his direction. As the decade moved along, the idea of creating a separate section to perform these calculations began to creep into field-artillery doctrine despite much opposition.[74] The proposed changes meant that when the forward observer later assumed the responsibility for conducting fire, he would not be compelled to make mathematical calculations on the spot during the heat and confusion of combat.

In 1929 Maj. Carlos Brewer, head of the Gunnery Department, discovered a book on the shelf in the school library about British artillery officer Neil Fraser-Tytler's experiences during the Great War, *Field Guns in France*. The author described how he had established observation posts at the very foremost positions and adjusted fire using map coordinates with very effective results.[75] Brewer discussed the success Fraser-Tytler had experienced with his new method to his fourteen colleagues in the department, and soon they too had read the book.[76]

Fraser-Tytler's placement of the artillery spotter led to a revolution in forward observation, for rather than adjusting fire from the traditional position of the battery commander at a post some distance from the front lines, the observer directed fire from a point where he could normally see the burst of the shells. The British officer further described how he had run a telephone line from his battery to the very front. From there he was able to see not only his target but also where the fire he was directing had landed. After word of his initial success of directing fire from the front lines had spread, additional telephone lines were strung out to his position, putting him in contact with several additional batteries, which enabled him to mass fires on the Germans very successfully in 1916.[77]

Taking away the battery commander's authority to command and adjust fire and giving it to the forward observer represented a monumental shift in command and control of artillery fire. Using Fraser-Tytler's idea of positioning the observer close enough to see the burst of shells to adjust fire on targets greatly enhanced field artillery's ability to provide responsive fire support that could meet battlefield conditions as they developed. Yet the new system that emerged needed further adjustments before it would reach the full extent of its capabilities to achieve true combined-arms effectiveness.

In its earliest stages, the new observation system linked one observer to a single battery. He could mass the fires of multiple batteries or conceivably an entire battalion on a single target if other observers could use his map coordinates to adjust their batteries on the target. This was a major improvement over the old system, greatly enhancing field artillery's ability to provide spontaneous fire support, but it still had two major drawbacks. First, because each battery computed its own firing data, it severely slowed the whole process. Also, the ability to mass fires was lost if the observer engaging the target was unable to contact other batteries' observers to pass along the grid coordinates of the target.[78]

By 1931 the Field Artillery School had developed a method for massing fires based on surveying the location of all batteries relative to each other and plotting them on a chart. Then, after one battery had adjusted fire, the target was plotted using the adjusting range and deflection. That point was then used to compute the data for the nonadjusted batteries. This chart was the forerunner of the observed-fire chart.[79]

The observed-fire chart in turn stimulated two other innovations. Another gunnery instructor, Lt. Charles Blanchard, developed a way to locate targets and firing batteries on a 1:20,000-grid map sheet. He then created a crude range-deflection protractor (RDP) for the grid sheet that the forward observer could use to quickly measure distance and direction from the gun to the target.

By the time Major Ward had replaced Brewer as head of the Gunnery Department in 1932, Brewer had developed a system similar to Fraser-Tytler's. Under the new system, the observer would still have to compute the firing data for the guns from his position at the front, using the map chart, firing tables, and RDP.[80] Brewer may have realized this was impractical, working on his own idea for solving the problem before leaving the department. In any case, the creation of the fire-direction center (FDC) apparently began as early 1931, but the term did not appear in any official publication until 1935.[81]

Regardless of what stage of development this method was in when Ward stepped in, he made it a reality by instructing that all computations be done in a sheltered FDC. The battalion FDC would be the origin of all firing computations. While the placement of the forward observer gave field artillery the ability to provide responsive fire on targets as they appeared, the creation of the FDC, Blanchard's chart, and the RDP gave artillery the capability of shifting fires in a timely and accurate manner through the forward observer.[82]

In the spring of 1932, Ward conducted an exercise at Fort Sill to adjust fire from individual guns in batteries and to mass the fires from an entire battalion on selected targets. Acting as a forward observer and using his own improved firing-calculation methods, Ward could adjust and mass fires so accurately and with such relative ease that he "compared it to 'squirting a hose.'"[83] Lieutenant Blanchard described a similar mass-fire exercise directed by a single forward observer that took place two months later, noting that the bombardment reached the target within eight minutes after the initial call. Over the next ten years, U.S. field artillery would cut that response time nearly in half.[84]

These new developments in early 1932 marked a major turning point in field artillery's ability to conduct more-responsive and effective fire. In fact, one former artilleryman noted that "with the introduction of these new fire direction techniques, the artillery made what can fairly be described as a quantum leap in its ability to participate in mobile warfare."[85] One might think that high-ranking officers would have applauded these improvements and welcomed the changes they would bring. But that was not the case. Many experienced, senior artillerymen resisted the new methods and disliked using the FDC. In particular, Maj. Gen. Upton Birnie, chief of field artillery from 1934 to 1938, strongly opposed the new system because it took the firing prerogative away from the battery commander and gave it to the forward observer.[86] So for several years, two departments of the same school gave their students conflicting doctrinal philosophies regarding responsibility for conduct of fire: one reassigned it to the forward observer, while the other insisted that it should remain with the battery commander.[87]

In 1939 Capt. John J. Burns gave a boost to the impetus for doctrinal change, writing that utilizing the battery commander to conduct fires was now obsolete. He argued that the battalion was a more effective fire unit than a single battery because as long as the battery commander initiated his own firing missions, it precluded two different batteries from simultaneously striking at the same target. Burns recommended using the FDC.[88]

FDC techniques that had been part of the field-artillery curriculum since 1934 did not begin to gain acceptance until the end of the decade. By that time, the turnover in senior personnel due to aging and retirement offered opportunities for promising new ideas and methods to gain acceptance. Thus, junior-grade officers who had received their field-artillery training in the 1930s promoted the use of the FDC as doctrine as they gained rank during the decade. In 1941 Gen. George C. Marshall, the army's chief of staff, settled the issue after observing a demonstration of massed fire by a division at Fort Sill. He almost immediately ordered Maj. Gen. Robert Danford, chief of field artillery, to implement the new system as standard practice. By February 1942 the army had made a landmark change by incorporating the FDC into its official doctrine by way of Field Manual 6–40, *Firing*.[89] This was the same year the artillery battalion table of organization first included the FDC under the S-3 battalion-operations section.[90]

The battalion FDC solved a number of problems simultaneously, eliminating the party-line system in communications and the resulting bottleneck that had hampered artillery's ability to fire multiple batteries simultaneously on a single target. It also transformed the definition of a forward observer from a somewhat generic term to something specific, while the new role created for the forward observer gave him many of the battery commander's former responsibilities. When working properly, the newly developed field radio greatly improved artillery's ability to communicate consistently.

Over a period of eight decades, radical technological improvements in artillery and small arms generally outpaced the doctrinal and tactical changes necessary to adapt to using these new weapons in combat. During the Civil War, the capability of artillery shells to carry for miles, well out of sight of the gunners, would eventually lead to a situation where observing the fall of shots was more important than aiming them visually, creating the need for a forward observer. As John Henry Grate, the last surviving Civil War veteran from Ohio noted in 1949: "In our time, we had to see the enemy to kill him."[91]

From that time on, increases in range and accuracy of small arms switched the tactical advantage from the offense to the defense and eventually led to the use of indirect artillery fire out of the necessity to protect the gun crews.[92] As noted earlier, J. B. A. Bailey described indirect fire as field artillery's most significant innovation in five centuries. Chris Bellamy elaborated on this idea, noting that it was pointless to have a long-range gun if its effectiveness was limited to the human eye and what was in its view. "The introduction of the modern system of indirect fire . . . was arguably the single most critical development in the history of land warfare of this period."[93]

Although U.S. field artillery in World War I was both deadly and effective, lack of observed fire limited its lethality and usefulness to situations involving preplanned fire and set-piece operations. Describing infantry assaults, Maj. Gen. Charles P. Summerall noted, "Infantrymen had no way to shift or concentrate firepower on unexpected areas of resistance, or to stop or start a preplanned barrage to keep pace with the momentum of an attack."[94] Doctrine and tactics had still not adapted to stay abreast of technology, and mobile warfare did not have much of a role to play until the latter stages of that war.

During the interwar period, several new developments created improved levels of command, control, and communications to benefit artillery's performance. The new system that removed conduct of fire from the battery commander and gave it to the forward observer through the FDC decentralized artillery's command and control. To maintain observed fire required a practical system of communication between the observer and the guns. The mechanization of field artillery that developed during the interwar years would facilitate the forward observer's ability to take wire longer distances into the field and lengthen telephone lines. The advent of practical field radios filled a significant gap in communications, although telephone wires would still be used and would remain vulnerable to shell fire since portable radios were not yet consistently reliable.[95]

On the eve of the American entrance into World War II, U.S. field artillery had developed new weaponry, equipment, doctrine, and tactics enabling it to mass the fires of multiple battalions on a single target quickly through the use of observers both on the ground and in the air. Despite having proven its ability to do this in peacetime exercises, no one knew with certainty how well this new system would work under fire. After all, respondents to various boards convened after World War I had expressed concerns about placing battery commanders, liaison officers, or forward observers too close to the front lines for fear that they would too quickly become casualties or that they would be incapable of directing fire amid the chaos of combat. Famed Russian artillerist Karl G. Guk, who died in 1910, predicted correctly that hiding the firing batteries would make the forward observers who directed their fire "prime targets."[96] In the interwar period, many even expressed doubts about the practicality of using aerial observation because the little planes used lacked speed and were unarmed. What these men would soon find, however, was that even though many forward observers would indeed become casualties, field artillery's success using the new system would surpass anyone's greatest expectations.

CHAPTER 2

Mobilizing for War

The 37th Division (Ohio National
Guard) and the 87th Division

The 37th Division (Ohio National Guard) and the 87th Infantry
Division were, in most respects, two typical American divisions. In
the Pacific theater of war, the 37th fought in campaigns on New
Georgia, Bougainville, and the Philippines. In Europe the 87th
saw combat in the Saar Valley, the Ardennes, and the Rhineland.
The artillery of both commands, one fighting the Japanese and the
other the Germans, may be regarded as broadly representative of all
American artillery usage during the war.

Nevertheless, both had some unique features too. In the 1930s
the 135th Field Artillery Battalion, Ohio National Guard was among
the first units to replace horses with trucks to move artillery. This
switch had huge implications for forward observers during World
War II. Another notable feature of the 37th Division—with particular application to forward observers—was Maj. Gen. Robert W.
Beightler's inclination to rely heavily on his field artillery. Although
he was certainly not the only general to follow this practice, Beightler often spoke of how he believed that using artillery had spared
the lives of many of his infantrymen.[1] The 37th along with the 1st
Cavalry Division were the only two U.S. army divisions to engage in
combat in a major urban population in the Pacific.

Brig. Gen. William Wallace Ford commanded the 87th Division's
artillery from March 13, 1944, until the division's deactivation. As a
young colonel, Ford was instrumental in convincing the U.S. Army

that light, single-engine aircraft could be used for aerial observation to direct artillery strikes. He also became the first officer to earn, and the only general to wear, an artillery-liaison pilot's wings. Elements of the 87th also breached portions of Germany's Westwall twice, first in December 1944 then again in February 1945; few American divisions could make that claim.

For three years prior to the outbreak of war in Europe, Chief of Staff Malin Craig had tried to streamline the army's combat organizations. One of the most significant organizational changes made took place in September 1939, when Craig ordered a restructuring of the army's infantry divisions from the "square" to the "triangular" setup. The standard mode of organization for expeditionary forces in World War I was square infantry divisions, so-called because they were two by two, that is, two infantry brigades, each controlling two infantry regiments, with three field-artillery regiments within an artillery brigade. At full strength each regiment consisted of nearly 4,000 enlisted men and officers and included three battalions and a machine-gun company. (By comparison, during the Civil War, the size of a typical Union volunteer regiment consisted of about 1,000 total personnel.) Each battalion at full strength had about 1,000 enlisted men and officers. With support troops included, the total strength of a square division stood at about 40,000 men.[2]

The triangular division comprised three rifle companies in each of three battalions in each of three infantry regiments; three battalions of field artillery, each with three firing batteries; and one battalion of 155-mm howitzers. This paring down in size facilitated mobility because it required much less road space than the square division, and this in turn enabled it to deploy more rapidly.[3] The increase in mobility offered many important advantages for both infantry and field artillery. With usable roads, several squads of riflemen could be squeezed into a single two-and-half-ton truck and transported to a starting point for maneuver in combat.

One of the first field-artillery units to experiment with motorization was the 135th Field Artillery, still organized as a regiment at that time. On August 18, 1935, units of the 135th conducted a motorized test march to Fort Knox, Kentucky. With a contingent of 103 motor vehicles towing twenty-four 75-mm guns, it made the long journey from northern Ohio to Fort Knox. One group started from Toledo and traveled south, while the battery from Youngstown

Organization of a Typical Triangular Infantry Division

Division Headquarters General & Special Staff			
Infantry Regiment	Rifle Battalion	Rifle Battalion	Rifle Battalion
Infantry Regiment	Rifle Battalion	Rifle Battalion	Rifle Battalion
Infantry Regiment	Rifle Battalion	Rifle Battalion	Rifle Battalion
Division Artillery	Artillery Battalion 105 mm	Artillery Battalion 105 mm	Artillery Battalion 105 mm
Engineer Battalion		Air Section*	Artillery Battalion 155 mm
Medical Battalion			
Special Troops			
Signal Company			
Reconnaissance Troop			
Quarter Master Company			
Ordnance Company			
Military Police Platoon			
Division Band			

*Edgar Raines indicates that a standard triangular infantry division had five air sections, one for each firing battalion plus one for division artillery headquarters. The U.S. Army however, was slow to revise existing Field Artillery tables of organization and equipment to show their existence. Raines, *Eyes of the Artillery*, 130–31.

moved diagonally across the state, both columns picking up additional units as they progressed. The Youngstown battery had the longest distance to travel, 460 miles. The two lines converged at Patterson Field, just north of Dayton, where they camped Sunday night. The next morning the entire regiment drove on to Fort Knox. Fortunately, there were no mechanical breakdowns en route.[4]

The return trip began on Saturday, August 31. Once past Cincinnati, each battery commander was allowed to select the shortest route from there to reach his home station. Some chose to cover the

37th Division Motor Convoy, Camp Shelby, Miss., 1941. Photo courtesy of Donald L. Walker.

entire distance home in one day.[5] One local newspaper, the *Alliance Review,* reported on September 3 that "Battery C and other units in the 135th Field Artillery demonstrated . . . that motor equipment has definitely arrived as a method of military combat."[6] The article goes on to say that the Alliance unit made the 400-mile trip from Fort Knox in about twenty hours and that until this year, the regiment had been using horse-drawn vehicles.

In 1938 the 37th still followed the old system of organization, using the square division with a single field-artillery brigade, the 62nd, which included the 134th, 135th, and 136th Field Artillery Regiments. The 134th Regiment was primarily from the cities of southeastern Ohio, the 136th was generally from southwestern Ohio, while the 135th came principally from northern Ohio. About a third of the members of the 135th were from Stark County.[7] In 1939 that regiment celebrated its centennial year, with the *Field Artillery*

Journal noting that "the 135th FA and its sister regiment, the 134th FA, constitute the oldest National Guard Regiments outside of the original thirteen states."[8]

Germany's swift conquest of France in May 1940 prompted concern in Washington that the Nazis might also begin taking over French colonial possessions in Africa and use them as a spring-board for invading South America.[9] The American public reacted to the fall of France with increasing alarm over the nation's military preparedness. In turn, Congress began making tentative plans to mobilize the National Guard. On September 16, 1940, President Roosevelt federalized the first installment of the guard. That same day the Selective Training and Service Act of 1940 (also known as the Burke-Wadsworth Act) became law, the first peacetime draft in the nation's history.[10]

Entry into federal service for the second wave of guard units came on October 15 and included the 37th Division. All these units

Battery C, 135th Field Artillery, Ohio National Guard, Camp McCoy, Wisc., 1940.

were to report to Camp Shelby (near Hattiesburg, Mississippi), which was still under construction when the first elements of the division arrived on October 21. The first of nearly 10,000 draftees assigned to the unit began arriving at Shelby on January 22, 1941. These men would bring the division up to its full strength of 18,000 troops. To help ease resistance to the draft, the War Department assigned only Ohio draftees to the 37th.[11]

The success the Wehrmacht continued to enjoy in Europe during the summer of 1940 made the call to active duty more palatable for most of the guardsmen. The atmosphere of crisis Nazi Germany had created worldwide convinced many of them that mobilization had not been unnecessary. Yet by the spring of 1941, it seemed less likely that Britain was about to fall, and by June a major morale crisis had developed as most of the young men became anxious to return to their homes and families.[12] Of course, this discontent was not limited to the Ohioans, but "by June, the letters O H I O began appearing on latrine walls, artillery pieces, and cars." This supposedly meant "Over the Hill in October" and implied that the men would leave en masse if their federal service exceeded a full year. *Life* even published an article in August titled "This Is What the Soldiers Complain About," based on a reporter's interviews with members of an unnamed National Guard unit from the North training in the South. General Beightler wrote a letter to the magazine's editors insisting that most of the story had little application to his division. In September *Life* printed Beightler's letter with an editorial note that the 37th Division was not the article's subject.[13]

Finally, the Japanese attack on Pearl Harbor on December 7, 1941, removed all hope of early release from federal service. Even at that date the 37th was still organized as a square division.[14] On December 9 Beightler spoke to his officers and noncommissioned officers, warning that "the division, already well along in its training, would likely 'be in the vanguard of our forces when more of the Army is called upon to take [an] active part.'"[15] In time, the truth of the general's word were fulfilled when, in April 1943, the first elements of the 37th Division arrived on Guadalcanal.

By the end of 1941, the field artillery was still searching for the right airplane to be used for aerial observation and trying to authorize their permanent attachment to infantry divisions; the Army Air Corps strongly opposed this measure.[16] Since the late 1930s, the

Entrance to Camp Shelby, Miss., 1941. Photo courtesy of Donald L. Walker.

army had been using the North American O-47, a large, bulky plane that required a relatively long takeoff area. Although the plane could reach a top speed of 221 miles per hour, it had poor maneuverability, further reducing its suitability for aerial observation.[17] In 1936 two officers of the Texas National Guard, Lt. Joseph F. Watson, Jr., and Capt. George K. Burr, who owned their own light, single-engine planes began using them to direct artillery fire during their unit's summer training exercises. Meanwhile, the development of early prototype helicopters siphoned away funds available to the air corps.[18]

Even as the problem of finding the right observation aircraft was under review, the field artillery helped itself by taking the lead among the ground arms in seeking a solution to the problem. Fortunately, the chief of field artillery, Major General Danford, also liked the idea of assigning planes directly to field units rather than to corps headquarters. In addition, he favored allowing field artillery to develop its own system for aerial observation.[19]

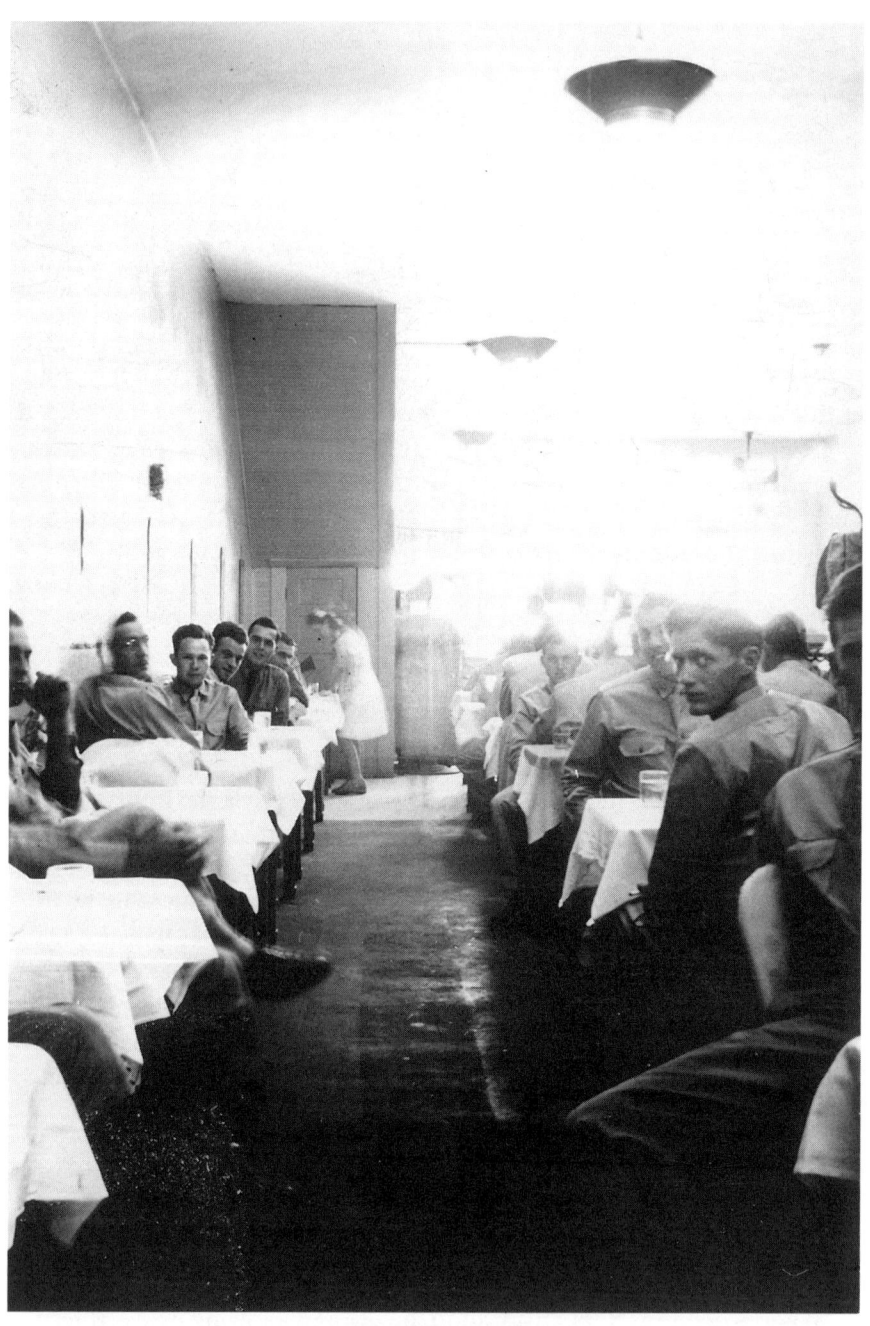

Members of Battery C in Doris's Coffee Shop, Hattiesburg, Miss., 1941. Photo courtesy of Donald L. Walker.

During the 1940 maneuvers held at Camp Beauregard, Louisiana, William Ford, then a captain commanding Battery I, 1st Artillery, expressed his dissatisfaction because of the difficulty he and other officers had in finding an air-corps observer to perform aerial spotting for their units. Lt. Col. John B. Wogan of the 68th Regiment wrote an article, published in the February 1941 edition of *Field Artillery Journal*, arguing for the creation of organic aviation units within the artillery for the purpose of observation. Wogan offered many sound reasons for his case, writing, "the artillery handles its own communications and its own supply of ammunition, but for some unfathomable reason, it must keep its hand off its air observation, which is just as vital to its proper functioning as are its communications and ammunition." He added that during World War I, reconnaissance planes had not maintained contact with supporting artillery units at crucial times, noting: "This arrangement has been a miserable failure in the past, and there is nothing to indicate its success in the future."[20] Wogan then declared: "It has been repeatedly and emphatically stated by many field artillery officers that they did not witness or hear of a successful air-ground artillery mission throughout all the maneuvers of the past years. This condition will not change as long as we must depend upon another agency for air observers and observation planes."[21]

Prior to any actual use in combat, the Army Air Corps objected to the idea of relying on unarmed observation planes to direct artillery fire because they thought that enemy fighters would simply shoot them out of the skies. Ford addressed this issue directly, writing that, like Colonel Wogan and General Danford, "he wanted both pilots and observers drawn from the Field Artillery." Then he took Wogan's argument one step further, writing that if it were necessary to gain air superiority or even air parity before landing ground forces on the European continent, than there would be a place for light aircraft on the battlefield.[22] Largely through Ford's efforts, by June 1942 the field artillery had obtained organic aviation units after a three-and-half-year struggle. That same year Ford attained the rank of lieutenant colonel.[23]

By July 1942 the U.S. Army had selected three companies to build the frames of the observation planes and one company, Continental, to build the engines.[24] The models to be used were designated the L-2, L-3, and L-4. The L-2 was the fastest of the three,

Lt. Clemens Schmitt, 334th Field Artillery, in front of L-4 plane near Saalfeld, Germany, May 1945. Photo courtesy of Donald L. Walker.

though it was also the most dangerous: at slow speeds it could stall without warning. The L-4 was much better suited for short, bumpy airstrips because it could fly slower and required less distance to take off or land.[25] While the L-4 was not perfect, in Ford's opinion it was the safest for use in the field.[26] Artillerymen referred to all three planes generically as "Grasshoppers." Ford spent most of 1943 with the Department of Air Training until October 23, when he left on temporary duty. By March 1944 he had become commander of the 87th Division Artillery.[27]

Maybe there was an advantage in sending National Guard units into combat before regular-army divisions during World War II. Although the guard typically had lacked adequate funding in the interwar years, it nonetheless underwent continual training during that time. For National Guard field-artillery units, this meant working with the French 75-mm howitzers left over from World War I. Nevertheless, in the last years before the war, guardsmen gained valuable skills and expertise from their training and had much more time to work and train together as a team than their regular-army counterparts.[28]

During World War I, the 87th deployed to France, and "in November 1918 when the war ended, the division was on orders to move to the front. In 1921 the 87th Division was reconstituted into the Organized Reserves and allotted to the Fourth Corps area. It was reorganized as a triangular division in 1942. Nearly two years later, December 15, 1943 to be exact—the 87th 'Acorn' Infantry Division was officially made an active division at Camp McCain, Mississippi."[29] Surviving veterans of Battery A, 334th Field Artillery Battalion recall that between the time of arrival at McCain and their departure for Europe, most of the original permanent cadre were gone.[30] Thus, in contrast to their counterparts in the Ohio National Guard, they experienced a much briefer time to train together. The 87th's "'Final Preparation for Combat' began on 21 February 1944," at Fort Jackson, South Carolina. "On 5 December the Division began its movement to the combat area in the vicinity of Metz."[31] By then the 37th Division had wrapped up its first campaigns on New Georgia and was in the initial stages of its second on Bougainville and was preparing to invade Luzon.

While neither division, the 37th or the 87th, would gain a reputation as a workhorse in its respective theater, each would experience heavy fighting and would perform admirable service. Forward observers, in their designated role as technicians, would be the key to their artillery's ability to provide close support in coordinating their efforts with their infantry. No one knew yet how the new system of assigning forward-observer parties to accompany maneuvering infantry would work out. But not only would it work well, artillerymen in the front lines would do much more than simply carry out their assigned duties.

| # Tactical Parity˙ and Disparity

Different Enemies

Parity, similarity, or relative equality in warfare can exist between opposing forces in a multitude of different military facets—technology, training, or logistics—or at the levels of tactical or strategic doctrine.[1] On the home front, these factors may include manpower mobilization, natural resources, and industrial capacity. Additionally, adversaries may be evenly matched in some areas, moderately matched in some, and very unevenly matched in others. What is worth noting is that in ground combat during World War II, there was a large overall disparity in nearly all of these things between the Japanese and the Americans, while in the ground war against Germany, relative parity existed.

The German army was much more advanced and experienced in the practice of combined-arms warfare than the Imperial Japanese Army. Part of this success was due to its ability to coordinate effectively artillery support of infantry attacks, which, in turn, was largely due to their use of forward observers.[2] In fact, with few exceptions, German artillery doctrine was very similar to that of the United States.[3]

In contrast, the Japanese were not nearly as far developed in the coordinated use of combat arms on the battlefield. This was due largely to their faith in the superiority of their own infantrymen.[4] Japanese tactical doctrine put great emphasis on three factors: offensive actions, surprise, and rapidity of movement. The foundation of this doctrine was the idea that a simple plan, executed with power and determination in combination with speed and maneuver, will upset the plans of enemy forces and prove successful. Japanese field-service regulations stressed, "the object of all maneuver is to close quickly with the enemy."[5] Their commanders believed that the sheer determination of soldiers could almost single-handedly win battles. This is what Allan Millett and Williamson Murray have termed "a reliance on the notions of spiritual strength and cold steel . . . to make up for material efficiencies." Underlying all operations was the idea of "faith in certain victory."[6] The lives of individual Japanese soldiers were only important as a means of service to the emperor.[7]

The Russians as well as the Japanese frequently used human-wave attacks, only in their case the attackers often had guns at their backs; Japanese infantryman went voluntarily. Yet the banzai charge was a self-defeating tactic.[8] British field marshal William J. Slim believed that Japanese commanders suffered from what some have called "Victory Disease." Based on Japan's triumphs early in the war, they had "an unquenchable military optimism" that almost never made any provision for setbacks or delays. National pride and individual vanity, therefore, played a large role in the inadequacies of the Imperial Army's doctrine and tactics.[9]

Although many American GIs believed that the Japanese fought fanatically, most admitted they were brave soldiers and tough fighters. John Stannard, a member of the Americal Division, called the Japanese "first class infantry soldiers, brave, tough, strong, patient, dedicated, obedient, loyal. They were the best."[10] Even though sound military doctrine was sorely lacking throughout the war, the spirit of the individual soldier made the Japanese army a difficult foe to defeat.[11]

In contrast, most American soldiers in Europe did not consider the Germans to be nearly as zealous, with the exception of some SS troops.[12] By September 1939 Germany had developed a fairly sophisticated combined-arms system of mobile warfare. House writes that during the May 1940 invasion of Belgium, German training in combined-arms tactics was particularly obvious during the sweep through the Ardennes. Combat engineers made possible a rapid advance over a poor road network, while antiaircraft guns played a key role in warding off Allied air attacks.[13] Although the German army was ahead of the U.S. Army in its development of combined-arms tactics, after entering the war, the Americans soon caught up.[14] In terms of tactical doctrine, then, relative parity between the German and American armies existed but not between the Japanese and Americans.

Technology, weaponry, and logistics also played important roles in the matchup of opposing armies. Here the same pattern of disparity with Japan and similarity with Germany was apparent. The Japanese army was particularly weak in motorization and mechanization. During the early years of the war in China, foreign observers reported that Japan was attempting to motorize its artillery, yet throughout the war most of it remained horse drawn.[15] The weapons of the Japanese infantryman were not up to par with those of the United States and Europe. The standard-issue rifle was the Model 1905 Arisaka, a Mauser bolt-action .256-caliber shoulder arm. Not only was the Arisaka difficult to fire rapidly, but it also had a limited range and accuracy.[16] The army had neither effective armor support for its infantrymen nor an effective antitank gun to combat American tanks.[17]

If Japanese weaponry and technology were lacking, their logistics were even worse. The struggle for Guadalcanal has been called a battle of logistics, one that the Japanese lost, setting the pattern for the rest of the Pacific War. Writing in 1991 in his introduction to a reprint of *Handbook on Japanese Military Forces,* David Isby observed that by 1944 the Japanese were paying a price for neglecting logistics and communications: "They saw themselves as warriors, conquerors, and samurai. The job of getting the men their food and equipment was not what a warrior did. The fact that the economy could never produce enough food or materiel certainly discouraged them from looking to these elements as deliverance from their strategic and tactical problems."[18] The ordinary foot soldier received a substantially less amount of daily rations than his American counterpart did.[19] Stanley Frankel and others who have given historical accounts of fighting on the Pacific islands frequently mentioned how the few Japanese prisoners the Americans did capture were frequently starving.

A comparison of Japanese and U.S. field artillery reveals the same pattern of inequality. Japanese field guns were generally lighter than their Western counterparts, although the projectile weight per caliber was about the same. As a result, the projectile velocity was slower than comparable guns of Europe and the United States, leaving Japanese guns with less range and power. In addition, their field guns characteristically lacked sophisticated sighting, consistent accuracy, and adequate variety of ammunition. Japanese artillery doctrine made poor use of counterbattery fire and employed inefficient tactics.[20] Brig. Gen. Harold Barker, who attributed the 136th Field Artillery with saving his life on New Georgia, classified Japanese counterbattery fire as generally poor, either because their forces lacked sufficient medium-range guns or because they were unable to master the basic principles of artillery.[21]

Reasons for poor Japanese counterbattery fire vary. First, during the early fighting in China, there was little need for it. Chinese artillery was so ineffective that the Japanese even placed their batteries in the open, with the result that their artillerymen suffered as high a casualty rate as their infantry.[22] Also, as the Japanese infantry attacked, the artillery moved its batteries forward to within 500–800 yards of the line of departure. This enabled them to support the main focus of the attack without moving but potentially exposed them to enemy counterbattery fire.[23] Last, among the fundamental principles of Japanese artillery tactics was firing in mass. Yet no specific missions were assigned, such as the support of particular infantry units. The commander of division artillery controlled direct support of the assault as well as counterbattery fire.[24] Such centralized control limited the artillery's ability to provide effective counterbattery fire as needed.

In the latter stages of the war, chronic shortages of ammunition hampered the Japanese. Even though these undoubtedly reduced the overall effectiveness of their artillery, the Americans could not disregard enemy batteries. As their ammunition became short, Japanese artillerymen would often fire a few intermittent shots, giving the impression of mere harassing fire, before hitting the Americans with a heavy barrage.[25]

In sum, Japanese artillery was not the equal of its American counterpart in tactical doctrine, equipment, or mobility. It never achieved the same degree of coordination with its infantry. Also, the Japanese placed less reliance on forward observers on the ground than did the U.S. Army. As a result it used less observed fire and attained poorer results.[26]

A comparison of German to American artillery shows much closer parity. German field pieces were on par with, and in some cases superior to, American field guns. They might have been even better except that, at the beginning of the war, the nation's high command believed that hostilities would not last very long and thus ruled out continuing research on promising but unproven artillery weapons anticipated to take more than a year to develop. Thus, they abandoned many projects that might have proven worthwhile. By the time the leadership lifted this ban a few years later, they had stretched their nation's resources too thin.[27] Mechanically, the 105-mm howitzer was almost identical with its American counterpart, but the Germans had not developed rapid fire-direction procedures or sufficient communications essential to massing fires quickly by battalion.[28]

The Wehrmacht's tactical use of its field artillery was good but not as effective as the Americans.[29] Like the Americans, they placed great emphasis on the importance of observed fires, using forward observers. A *Field Artillery Journal* article in December 1941 drew on German sources to describe what the Germans considered to be the most successful features of their field artillery. Despite any propaganda value it might have intended, the article attributed their artillery's success largely to excellent training and to relying heavily upon observed indirect fires.[30]

Further evidence of the importance of artillery observation appears in a translated German field manual in 1944. Regarding the use of observation posts, it states that infantry and artillery may at times share the same post, but the ranking officer present will decide priority and would generally allot to artillery the best positions, although he would still arrange for a "reliable intercommunication for exchange of observations." This reveals the high priority the German army typically gave its artillery over infantry in this regard.[31]

The army also placed great importance upon the use of counterbattery fire. General der Artillerie Karl Thoholte claimed that during three or four months of continuous and heavy defensive fighting, counterbattery fire knocked out 4,000 enemy batteries on the Eastern Front, adding that in instances where enemy artillery was the main factor affecting his ability to attack, counterbattery fire quickly became his units' primary mission. The general also indicated that he could not use counterbattery fire as extensively in the latter stages of the war on the Western Front because it took so much ammunition, then unavailable, a primary reason for this failure.[32]

A relative lack of field-artillery mobility proved to be a prime weakness. The Germans had not obtained sufficient motorized transport and depended heavily upon horse-drawn artillery. Petroleum was already in short supply when America entered the war and would become even more so as the war continued. So, instead, the Germans used horses to move most of the combat and field trains of the infantry by December 1941.[33] Commenting on this continuing dependence upon horses, General Thoholte noted that an important lesson he had learned from the Eastern Front was that artillery reinforcement needed to be as mobile as possible. He praised American artillery, calling it the most mobile of all first-rate powers, adding that the whole army was the most motorized in the world.[34]

Bruce Gudmundsson rates American artillery about even with its German counterpart, noting that comparative mobility was the biggest difference. This was because "save for a few batteries of pack howitzers carried by mules, all U.S. field artillery was either towed by motor vehicles or self-propelled." He also gives the United States the advantage in the ability to mass fires because the army generally maintained ample supplies of ammunition. But Gudmundsson rates German artillery higher in its ability to mass the fires of many divisions, attributing this to the less flexible nature of U.S. command arrangements. Whereas the Germans would consent to putting one division's artillery under the direct operational command of another, the Americans were reluctant to do this. The United States also kept its division and corps artillery discrete.[35]

Gudmundsson may have based these last observations on the comparative frequency with which it occurred. That is not to say, however, that U.S. field artillery was not capable of massing the fires of multiple divisions. During the Japanese attack on Hill 700 on Bougainville on March 24, 1944, all artillery of the 37th Division plus three battalions of the American Division fired 6,450 rounds of ammunition, completely halting the Japanese assault.[36] This may not have been quite the scale of firepower Gudmundsson had in mind, yet it did represent the massing of two divisions' fires. In Germany, in support of the 345th Infantry Regiment's crossing of the Mosel River on March 16, 1945, "nearly 300 guns (of VIII Corps) fired over 5,000 rounds in a five-minute barrage."[37] Although Gud-mundsson's observation may be correct with regard to instances involving multiple divisions, after the war Gen. George C. Marshall wrote: "We believe that our use of massed heavy artillery fire was far more effective than the German techniques and clearly outclassed the Japanese."[38]

An overall comparison of German to American artillery, then, shows that in terms of doctrine, tactics, and equipment, they were fairly evenly matched. Where German artillery fell short was with its mobility. The Wehrmacht's overdependence upon horses for transport affected not only its artillery's mobility but its all-important system of logistics as well. Without adequate logistical support, German gunners suffered not only from a lack of ammunition but also from shortages of food—for both men and horses—and medicine.

The comparison of opposing armies thus far has focused largely on the ways in which they differed, although the German and American armies certainly had more similarities than differences. To understand what made the American experience in the Pacific different than in Europe, secondary differences between the two theaters of combat must be studied. These are what might be called leveling factors since they affected both sides equally. In particular, these include climate, topography, and population density.

To begin with, climate had a tremendous influence on a variety of things, ranging from adverse effects on equipment to the health of the men. The obvious differences in climate included the tropical heat, humidity, and rainfall on the islands of the Pacific compared to the more temperate climate (like that of the northern United States) found across most of Western Europe. Aside from the discomfort of fighting in extreme heat and humidity, a host of diseases not encountered in Europe plagued soldiers fighting on the islands. At Okinawa Maj. Eugene R. Smyth emphasized the importance of health and sanitation issues because unless the strictest safeguards were maintained, the resulting diseases could quickly impair the combat efficiency of his men very seriously.

Smyth mentioned the distribution of Atabrine among the troops, and how his unit had spent fourteen months in the combat zone without a single case of malaria and only a few cases of dengue fever. He added that although prevalent diseases included amoebic dysentery and liver flukes, only one or two cases had occurred in his artillery battalion, usually among forward-observer parties after they had spent a few days in the swamps. The major emphasized that, although difficult to control, skin diseases were 99 percent preventable. Rats created a problem on Okinawa, but by 1945 the army began using C-47s to spray large land areas with DDT.[39]

Until the latter stages of the Pacific War, the army concentrated more on fighting the enemy than on controlling tropical diseases. Sgt. Don Allison and Tech Sgt. Abbie Cohan, members of Battery C, 135th Field Artillery Battalion, 37th Division, contracted malaria, while six others in the battery became sick with other unspecified illnesses after reaching the Pacific theater. Soldiers on Bougainville also encountered a variety of insects that plagued them routinely. Centipedes lived in their foxholes. Ants twice as large as black ants in the United States made themselves a particular nuisance, their bite feeling "like a stab from a hot needle." Maybe even worse was a thorny barbed vine that grew plentifully in both the trees and on the ground.[40]

Differences in climate between the Pacific and European theaters also meant a disparity between the number of noncombat deaths in the two theaters. In his summary of 37th Division casualties, Stanley Frankel lists 1,127 killed and 218 who later died of wounds, or a total of 1,345 men. Frankel's compilation indicates 73 members of the 37th who died from other causes.[41] This means that approximately 5 percent of the division's deaths came not as a result of battle. A reasonable conclusion, then, is that the majority of these noncombat deaths were the result of disease.

The incidence of fatal disease among U.S. soldiers was much greater in the Pacific than in the European theater. During the entire war, 26,518 Americans serving in all theaters died from diseases. Of these, 2,474 succumbed in Europe, while 10,828—more than four times as many—died in the Pacific.[42] The adjutant general's *Army Battle Casualties Final Report,* a compilation of all army casualties incurred in no less than twelve theaters, indicates that more than twice as many combat deaths occurred in the European theater (135,576) than in the Pacific (50,385).[43] Thus, in sheer totals, without any computation of what percentage of military personnel serving in each of these theaters died from these two causes, the gross number of battle deaths was four times higher in Europe than in the Pacific, while the number of fatalities due to disease was four times higher in the Pacific.

Of course, the adverse effects of fighting in a tropical climate did not apply to the Americans alone. As mentioned before, Japanese soldiers suffered much more seriously from poor diet and lack of medical care. An Americal Division operations memo for Bougainville reported: "Prisoners of war and Japanese bodies examined show that almost 90% of the 6th Division was suffering from malnutrition, malaria, dysentery, beri-beri, or skin diseases. Thus their fighting efficiency both physically and spiritually was greatly impaired."[44]

Although Americans fighting the Germans also died from causes unrelated to battle, the prevalence and severity of the diseases and, in particular, the mortality rate in Western Europe may not have been on the same scale as that of the Pacific theater. Living outdoors involves many unpleasantries in any area of war. Although Americans in the European theater escaped the heat and humidity that their counterparts in the Pacific endured year round, they nevertheless had to contend with other extremes such as living in a foxhole during the bitter cold of winter.

By definition, a casualty is a soldier lost to active service as a result of death, injury, or capture, so not every casualty is a fatality. During the war, air and ground troops suffered 91,000 cold-weather casualties, including frostbite, pneumonia, or other related ailments, 45,283 of these occurring between November 1944 and April 1945. This represents nearly half of all American air and ground cold-weather casualties during the entire conflict and about 9 percent of those incurred during the last six months of the war in Europe. On average, about 7.6 percent of all American soldiers in Europe, nearly one out of every fourteen, became casualties due to cold weather.[45] No comparable figure exists for the Pacific theater.

Differences in climate between the two theaters also affected the communications systems of the combatants significantly. The constant humidity of the Pacific islands decreased the reliability of battery-operated field radios, which often forced forward observers to rely on field telephones to communicate with their battalions. The *Observer's Checklist* for artillery officers departing for the Pacific reported that on Buna and Guadalcanal, artillerymen used wire almost exclusively. Had the 600-series radio equipment been available, however, the situation might have been different.[46] If field radios were inherently unreliable in the years of 1942 and 1943, the constant humidity of the jungles had a particularly adverse effect on their performance.

Unfortunately, artillerymen could not necessarily depend upon telephone lines on the islands either, though not because of the climate. Lt. Col. John W. Ferris described the problems encountered when relying upon them in the jungle, explaining that as long as the lines remained intact, they worked more reliably than contemporary portable radios. But telephones had problems of their own. Of course, the Japanese cut the lines whenever they found them, but American soldiers unintentionally caused many breaks. As the assault troops advanced, they hacked out a narrow trail through the trees and vegetation and strung their telephone lines along the sides. As these forward areas became secured, the support troops that followed widened these lanes to make space for vehicles, cutting down many trees along the way and, in the process, severing the telephone lines. Lack of manpower and time among the engineers prohibited trying to protect or even save these wires, and they cut them indiscriminately to save time. The only solution for artillery was to provide wire crews to work alongside the men on the bulldozers, stringing new telephone lines to replace those severed.[47]

In contrast, by 1944 in Europe, field radios seemed to have been much more reliable, in large part due to less precipitation and lower humidity. Yet another reason was that throughout the war, radio technology gradually improved. In France in the summer of that year, Lt. Col. Frank W. Norris acknowledged that the forward-observer parties had, for the most part, come to depend upon field radios, emphasizing his own complete faith in them. He felt that if properly maintained and by using proper transmitting discipline, reliance on radios presented no problems under normal circumstances in that part of the world.[48]

But the use of radios had two drawbacks that may have been more serious in Europe than in the Pacific. First, in any theater of the war, radio antennas were conspicuous enough that they tended to draw fire. John Colby, 90th Infantry Division historian, describes the experience of forward-observer parties with radios: "Unfortunately the radio had a long whip antenna that could be spotted easily by the enemy and often was. In our view, those radios attracted fire like magnets. Even more unfortunately, the attracted fire tended to eliminate the FOs as well as the man who carried the radio."[49] For example, David Thibodeau landed on Utah Beach on June 9, 1944, and served as a radio operator with Battery B, 42nd Field Artillery Battalion, 4th Infantry Division until he was captured five months later. Thibodeau noted that in mid-July at Mortain, he spent a nervous fifteen minutes while a German machine gunner kept raking bullets back and forth over his position and watched the man lying next to him bleed to death after being struck by a bullet in the throat. Thibodeau was sure the enemy gunner had seen the radio on his back.[50] In any case, forward observers, denoted by their binoculars and radios, were priority targets for German snipers and infantry.[51]

Second, the Germans were apparently more skilled than the Japanese at detecting the sources of American radio waves. Through the process of triangulation, they could locate the point of origin of a radio transmission, then use their own artillery to fire on it. In instances where American units suspected that this was happening, they either switched back to the field telephone or limited their use of radios.[52]

In addition to climate, the amount of physical, geographical space in which individual campaigns took place played a role in creating differences in how the ground war was fought in the Pacific and Europe. Ground combat in the Pacific was spread over an area encompassing thousands of square miles in total, yet each island campaign occurred within the confines of a relatively small land area. Space available to maneuver was limited, and land forces fought in relatively confined geography compared to the battles fought on the European continent. Among the largest Pacific battlefields were the islands of Luzon and Okinawa. The largest of the Ryukuan Islands, Okinawa, is approximately sixty miles long and forty-five miles across at its widest point. On the other extreme is Iwo Jima, a land mass of volcanic ash encompassing no more than sixteen square miles. In contrast, soldiers fighting in Europe had much more room to maneuver and engage. This was a significant factor affecting how the war was fought in both theaters.

Fighting an enemy in a relatively small, confined area meant that the artillery crews were typically in closer contact with enemy infantrymen than their counterparts in Europe. Although there are some very notable exceptions in Europe, American artillerymen in the Pacific were more likely to engage the enemy directly in ground combat at the battalion level. For example, Lt. Bruce Wells recalled an incident on Damulaan when a Japanese suicide party managed to infiltrate one of the batteries. As a result, artillerymen fought as infantrymen. In the brief firefight that ensured, one Japanese soldier made it beside the trail of a howitzer before he was killed, while a second one managed to detonate a heavy satchel charge in the breech block of another gun. By dawn, the artillerymen had killed all the enemy except one, who was severely wounded and taken prisoner.[53]

Topographical differences between the Pacific and European theaters affected field artillery in other ways. Artillery officer John Casey explained that on Guadalcanal it was often necessary to use high-angle fire due in part to the close proximity of the Japanese. Moreover, high-angle fire put shells on a trajectory that would enable them to hit enemy troops dug in to the reverse side of a slope or to avoid hitting treetops and exploding prematurely.[54] Capt. Ralph M. Fuller indicated that this was very useful for jungle warfare since most such operations took place only 500–1,500 yards from the batteries on the beachheads. He advised battalion commanders to position their guns as far from the units they supported as possible, recommending 2,800 yards, or about a mile and a half, as the minimum distance at which artillery could support infantry units with high-angle fire.[55]

Differences between the two theaters in plants, trees, and the density of foliage also made a difference in how the army chose to fight and use its artillery. As forward observers on New Georgia and Bougainville quickly learned, they would frequently have to use sound instead of sight to sense rounds and adjust fire since the thick tropical vegetation often hid the flash of bursting shells. From Okinawa, Maj. Eugene R. Smyth noted that the spotters in his battalion adjusted fire by sound at least three-fourths of the time in combat and became very proficient at it.[56] There was an added advantage in that it could also be used for spotting at night. Forward observers in Europe sometimes used sound when visibility was limited.

From his combat experience in France, Capt. Eugene Maurey came to realize that he was often close enough to enemy guns and mortar positions that he could hear their propelling charges and determine the distance and direction by ear. Elated by his success while serving with the 79th Division through the Foret de Parroy, Maurey wrote an article, published in *Field Artillery Journal,* describing his methods and encouraging other forward observers to try them as the circumstances allowed. He later published an account of his experiences as a forward observer during the war, one of only a few to do so.[57]

Comparatively less foliage in Europe meant fewer places for concealment, thus increasing the danger of gun positions drawing counterbattery fire. Muzzle flash and smoke typically gave away the Americans' positions. Capt. Richard Van Horne suggested using firing positions in the natural depressions of the ground.[58] One feature of the Normandy terrain that afforded natural concealment for American and German batteries alike was the hedgerow, providing natural protection as well as cover. The raised earth provided additional defilade, and the foliage helped conceal the muzzle blasts. Also it was easy to blend the camouflage in with the local terrain.[59] Here the forward observer with the ability to adjust fire by sound had a real advantage.

A third leveling factor of sorts between units fighting in the Pacific and Europe was population density. This made a significant difference in how the ground war was fought and, in turn, had implications for field artillery and forward observers. Although much of the combat in Europe took place in relatively rural areas, the continent had many more inhabitants per square mile than the majority of islands in the Pacific.

Population density affected the quality of maps available for general military use and particularly for artillery, which depended upon precise topographical maps to direct fire accurately. Prior to World War II, cartographers worldwide had placed more emphasis on Europe than the remote, sparsely populated islands of the Pacific. As a result, early reports of ground fighting in the Pacific indicated that the maps available were not entirely satisfactory. Commenting on the inaccuracies resulting from combining smaller, individual sections for artillery use on New Georgia, Lt. Col. Robert Gildart complained that the grids appearing on the maps made from aerial photos were rarely true squares and sometimes were not from the same altitude. Instead, the firing charts had been assembled by matching up details on photographs of adjacent areas, distorting both the scale and the azimuth. Consequently, a map of this kind was unsuitable for artillery use. Gildart did mention, however, that in addition to the unsatisfactory photomap, a multicolor map of the New Georgia area was available on a scale of 1:20,000 and that it was very satisfactory.[60]

Part of the problem with maps, too, was trying to locate reference points in a jungle. Gildart observed that on a map or photo showing only the natural features of a jungle, it is hard to determine the location of friendly troops with precision.[61] Here, the absence of any manmade landmarks probably made it difficult to distinguish one hill from another. An army bulletin summarizing the lessons learned from Bougainville stressed the importance of obtaining and making better maps before beginning an operation.[62]

In contrast, the maps available in Western Europe were noticeably more reliable. Lieutenant Colonel Norris praised the quality of his firing charts in France. He noted that both the survey section and the fire-direction center could use the same maps, thus increasing the number of personnel in the battery available for forward-observer parties.[63]

Perhaps, the greatest significance of population density was the concern for avoiding civilian casualties, what the U.S. Army now calls collateral damage. An estimated 100,000 civilians perished during the struggle to liberate Manila from the Japanese. During that battle, the 37th Division relied heavily upon its field artillery. Yet Manila, the largest urban battle the United States fought in the Pacific, was atypical of combat in that theater. Most of the islands where fighting took place were very sparsely populated. The more numerous urban centers and higher population density of Europe meant an increased concern for avoiding civilian casualties there. After the Rhine crossing, it was customary for the German citizens of small villages to hang white flags out their windows as the Americans approached as a sign that they would offer no resistance and a request to spare their town destruction by artillery.

More people in Europe meant more buildings, and because forward observers sought an elevated position to get the best view for directing fire, these structures gave them an advantage not normally found in the Pacific. Whenever they could, forward observers used towers and tall buildings, and because almost every little village had a church of some kind, steeples proved to be a common spot from which to direct artillery fire, for Germans and Americans alike. Because military observers of any kind are prime targets, artillery fire destroyed the steeples and damaged the churches in many villages throughout Western Europe, killing and wounding several spotters.[64]

For example, in the French village of La Hayed Puits, a German forward-observer crew set up an observation post in the steeple of the village cathedral. As the 79th Infantry Division approached the town, German infantrymen on the ground quickly left. The German observers lingered too long, however, because a round from the 312th Field Artillery Battalion made a direct hit on the steeple they occupied. Shortly after, when the riflemen of the 314th Infantry Regiment entered the village, they found the bodies of the Germans sprawled out on the public square.[65]

In a comparison of the U.S. Army's practice of combined-arms warfare during World War II to that of the Japanese and Germans, a notable imbalance existed in the Pacific theater and relative parity in the European theater. Japanese infantry never worked well in conjunction with its other combat arms, and in the case of artillery, the coordination was not particularly good. Germany's practice of combined arms was ahead of the Americans when they entered the war, but over time the U.S. Army caught up.

The Japanese predicated their military doctrine upon bold, swift attacks. The idea was that courage and spirit could make up for any material deficiencies, but in practice it had disastrous results. In effect, their tactical doctrine was mired somewhere between Gen. John J. Pershing's belief in the primacy of the rifleman and that of the armies of late-nineteenth-century Europe, one that cost them severely in World War I.[66] In contrast, the tactical doctrine of the Wehrmacht, while also emphasizing the importance of bold, aggressive action and especially mobility, recognized the advantage gained by using all available combat arms in concert. Japan's overall war effort was also severely handicapped because it lacked an abundance of natural resources, failed to maximize its industrial capacity, and placed little emphasis on a viable system of logistics. Fortunately for the world, Germany overextended itself by committing the same error it had in World War I, that of fighting on two fronts simultaneously.

Japanese field artillery was not the equal of its American counterpart. Japanese artillery doctrine followed a rigid chain of command that failed to provide for spontaneity and responsiveness. Their guns lacked the range and accuracy of the Americans, and ammunition shortages were commonplace. Japanese counterbattery fire was largely ineffective. But their optical equipment was superior to that used by the Americans, and U.S. field artillerymen used captured optics whenever they could get it.[67]

In contrast, German artillery doctrine and practice was quite similar to that of the United States. Their guns were generally on par with the Americans. Germany, however, lacked the ability to mass fires as effectively as the United States. The lack of motorization and continued dependence on horses severely hampered German mobility and, in turn, their overall effectiveness.

Factors regarding the physical conditions in the two theaters of the war, such as climate, topography, and population density, also played a role in the shaping of overall parity and disparity between the Americans and the Japanese and Germans. The climate of the Pacific theater was rough on both men and equipment. More American soldiers died from disease in the Pacific than in Europe. Humidity and dampness increased the unreliability of field radios in the Pacific. In Europe, though, cold temperatures would drain radio batteries quickly and sometimes freeze the operating mechanisms on various weapons at the most inopportune times. The climate also caused thousands of U.S. casualties due to frostbite. The topographical features of the jungle islands forced forward observers to rely often on sound rather than sight. In addition, fighting on an island made it more difficult to distance oneself from the enemy and sometimes forced field artillerymen to resort to using dangerous high-angle fire. Finally, because the population density in Europe surpassed that of the majority of combat zones in the Pacific, American soldiers in Europe typically had more difficulty trying to avoid killing civilians while engaged in combat.

Despite the tactical disparity in the Pacific and parity in Europe in ground combat, most American soldiers probably would have chosen to serve in the European rather than the Pacific theater of war. David Kaufman, a native of Cleveland, Ohio, trained with the 87th Division but was reassigned to another unit and ultimately served in the Pacific with the 129th Infantry Regiment, fighting the Japanese on Bougainville and Leyte. Years later he attended a reunion of the 87th Division so he could visit old friends he had known during training. During the gathering, the question came up as to which was tougher, fighting the Germans or fighting the Japanese. "The men at the 87th reunion said they all preferred the former not the latter. The Germans took prisoners."[68]

| # Baptism of Fire

The 37th Division on New Georgia

The participation of the 37th Division in the New Georgia campaign provided its artillerymen with their first opportunity to execute their new tactical doctrine using forward observers and the fire-direction center. This would take place in the hostile environment of a tropical island where the terrain and climate compounded the misery of fighting a tenacious and often invisible enemy. The forward-observation teams quickly learned that their greatest challenges centered around maintaining tactical control, observation, and communications. The need to be abreast, if not forward, of maneuvering infantry to observe the enemy and direct and adjust fires made their dangerous job even more hazardous. Terrain and climate made the use of heavy field radios difficult and often infeasible, while support troops struggling to perform their vital logistical functions at times inadvertently severed telephone lines.

New Georgia also gave the division's forward observers their first opportunity to put their training from Fort Sill to use. Prior to World War II, the term "forward observer" was a generic one. During World War I, it might have applied to a battery commander, an artillery-liaison officer, or any of the enlisted men assigned to help these two officers observe the fall of rounds. It was the battery commander, though, who conducted fire. Typically, he worked from an observation post to carry out his duties, although some

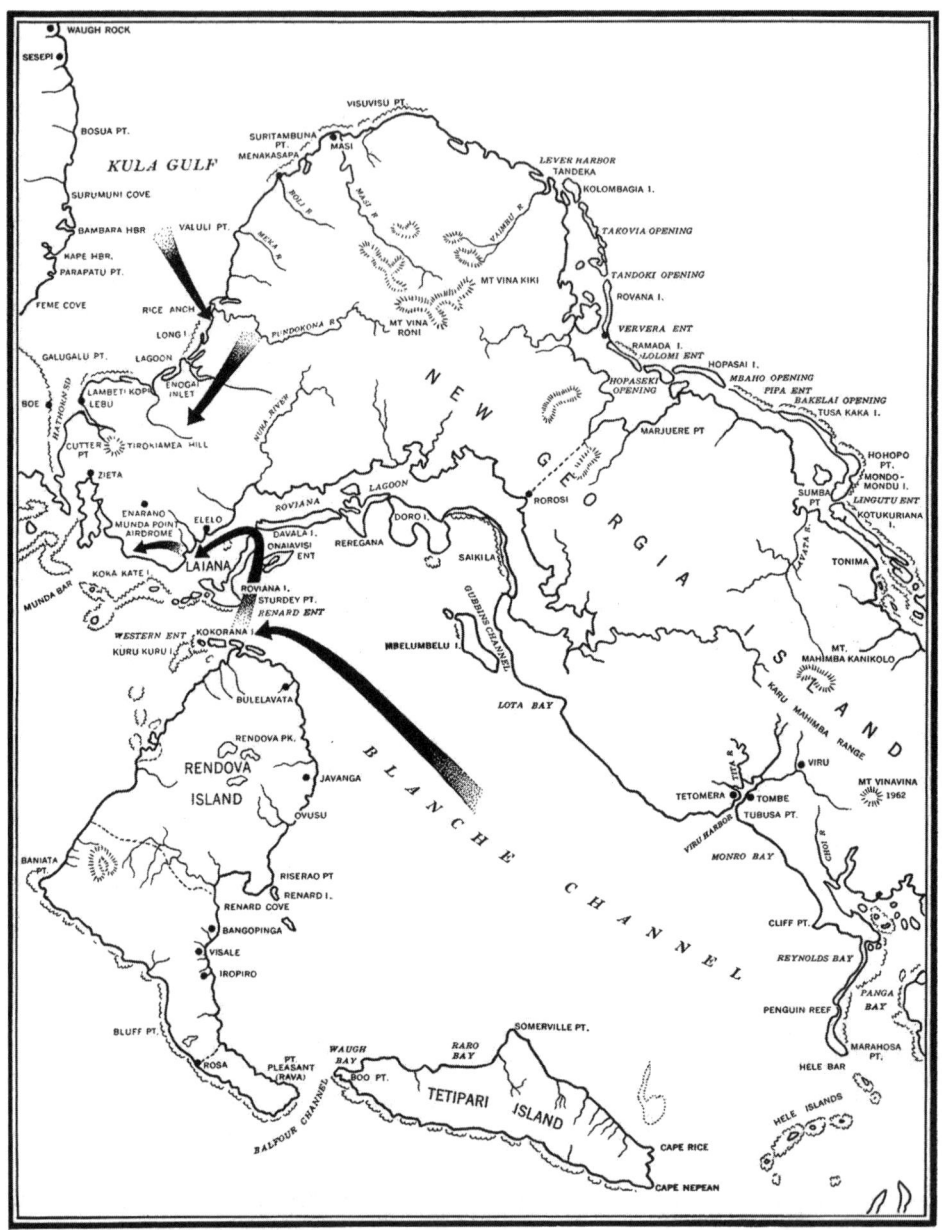

The 37th's plan of attack on New Georgia. From Stanley A. Frankel, *The 37th Infantry Division in World War II* (Washington, D.C.: Infantry Journal Press, 1948).

of the enlisted men stood very close to, if not in, the front lines with the infantrymen to help the commander adjust fire. If the riflemen managed to break through the enemy lines and continued their advance, the artillerymen remained behind. Not until World War II did forward observers stay with maneuvering infantry and control the conduct of fire.[1] As Capt. Conrad Boyle has noted regarding World War I, "No agency of communication existed at that time which would allow forward observers to exercise dependable control of the fire power available to them."[2]

By the time the U.S. Army committed ground forces during World War II, this had changed. Now the army designated an artillery officer, usually a second lieutenant, who had received training in forward-observation methods, to this role. He had a minimum of two enlisted artillerymen assigned to him to make up a forward-observation team. With at least one party per infantry battalion and more often one per rifle company, the group maintained perpetual contact with the infantry commander. Whether the riflemen advanced, retreated, or stayed in place, the observers remained with them until relieved. Although the frequency of relief varied somewhat, it was available to these frontline artillerymen more often than to combat infantrymen, perhaps leading the army to not formally recognize these men as combatants with a ribbon similar to the Combat Infantryman's Badge (CIB). But if one could talk to a forward-observer veteran of the 37th Division today, it would be difficult to convince him that the experiences he shared with riflemen while battling the Japanese did not represent actual combat.

In February 1942 the War Department directed the 37th to reorganize into a triangular division and later that same month ordered it to report to Indiantown Gap, Pennsylvania. Here, the men received even-more-intense training in preparation for overseas assignment.[3] True to General Beightler's prediction, the division was among the first in the army to deploy to a theater of war, but nearly a year would pass before the Buckeye Division first entered combat. By then its organization included the following elements: 129th Infantry Regiment; 145th Infantry Regiment; 148th Infantry Regiment; Headquarters and Headquarters Battery, Division Artillery; 6th Field Artillery Battalion (105-mm); 135th Field Artillery Battalion (105-mm); 136th Field Artillery Battalion (155-mm); 140th Field Artillery Battalion (105-mm); 37th Reconnaissance Troop

(Mechanized); 117th Engineer Combat Battalion; 112th Medical Battalion; 37th Counterintelligence Corps Detachment; Headquarters Special Troops, Headquarters Company, 37th Infantry Division; 737th Ordnance Light Maintenance Company; 37th Quartermaster Company; 37th Signal Company; and Military Police Platoon.[4]

Originally slated to go to Northern Ireland and then to New Zealand, a third change of orders assigned the Buckeye Division to the Fiji Islands.[5] On May 26, 1942, a convoy left San Francisco harbor carrying the 37th Division to its temporary assignment in the South Pacific. By mid-July the entire division had reached Fiji after a brief stopover in New Zealand.[6]

Ten months later the 37th was reassigned to Guadalcanal, arriving on April 6, 1943, six weeks after the fighting had ended. Stanley Frankel's history of the 37th Division does not record any casualties resulting from residual Japanese forces left on the island. But from the first day of their arrival, the enemy subjected the newcomers to air attacks.[7]

Capt. John Casey, an artillery officer who fought on Guadalcanal, offered a preview of what the 37th Division's artillerymen would soon experience. He observed that it was often necessary for crews to aim their guns at a much higher angle than normal to hit Japanese positions on the reverse slopes of open ridges because shells fired at the flatter trajectories typically used would either skim over, landing harmlessly beyond the enemy, or fall short, one of the cardinal sins a forward observer could not commit because of the possibility of friendly casualties. The flatter the trajectory of an artillery shell, the less likely it is to be pushed by the wind. Conversely, winds are more likely to affect where a shell lands when the round follows the abnormally higher trajectory Casey described. To use high-angle fire accurately demanded reduced tolerances in adjustment by forward observers than usual. Although its use had the desired effect, such fire was at times unpredictable and posed a risk to friendly troops. It could only be brought to within four hundred yards of American lines because its steep vertical plunge made it more susceptible to unpredictable drifting from prevailing winds. To use it safely and effectively depended upon close coordination between the observer adjusting the fire and the batteries and FDC.

Describing observation methods used for adjusting fire, Casey explained that only two methods were available, aerial and ground observation, and felt that the latter produced better results. Men on the ground carried the same basic equipment that the combat infantryman did in addition to the heavy portable field radios used to communicate with the FDC. For backup, they laid telephone wire as well.[8]

The humidity took a heavy toll on the batteries used in the field radios.[9] Although enemy artillery and mortar shells did sever some observers' telephone lines, the Japanese used their artillery sparingly on New Georgia.[10] During combat, forward observers needed a way to be able to discern exactly where shells were landing to avoid directing artillery fire on their own men. On the islands visual sighting was limited at times because mammoth trees and thick jungle growth often obscured the observers' view, not only during daylight but even at night, when the flash would normally be most visible in. So again, the ability to adjust artillery fire by sound became an important asset to many forward observers in the Pacific theater.

Guadalcanal, the first American land offensive of the Pacific War, proved false all prewar concerns about artillery's effectiveness in the jungle and the forward observers' ability to operate effectively from the front lines during combat. It also demonstrated the blurring of functions assigned to infantrymen and artillerymen that would occur wherever forward observers served. Casey noted that while accompanying the infantry: "We learned to travel light and quietly. Circumstances quickly taught us minor infantry tactics. We drank out of shell holes. We went for days without chow. The honeymoon was over. The artillery went to war."[11]

In most of the island campaigns, logistics played a crucial role in the final outcome. Guadalcanal and New Georgia were no exceptions. To offset their shortage of ammunition, Japanese artillerymen on New Georgia resorted to a ruse that frequently proved successful. Amid U.S. artillery barrages, they would sometimes lob in an occasional mortar or artillery shell on American positions in an effort to make the GIs and marines think that the exploding shells were their own artillery firing short. Often it worked. Immediately some concerned infantry lieutenant or captain would make an urgent request for the artillery to cease firing.[12]

Except for air raids, life on Guadalcanal for the men of the 37th Division, although primitive, was comparatively safe. Less than two hundred miles north lay the island of New Georgia, where Japanese aircraft from Munda Point began their bombing raids on American positions. By capturing Munda airfield, U.S. fighter planes could then escort bombers from Guadalcanal in raids on enemy positions on Bougainville and Rabaul, the largest Japanese naval installation in the South Pacific.

The plan of attack for the New Georgia campaign, code named Operation Toenails, called for U.S. Marines, joined by the army's 43rd Division, to carry out the initial assault while the 37th stood by on reserve. Beightler detached a number of units from his division, including the 3rd Battalions of the 145th and 148th Infantry and the 136th Artillery Battalion, to accompany the 1st Battalion, 1st Marine Raiders to land at Rice Anchorage on the northwest coast of New Georgia.[13]

The battle for the island lasted about a month. On July 5 the 3rd Battalion, 145th Infantry landed, followed by the marines and the 3rd Battalion, 148th Infantry. On July 9 the 136th Field Artillery fired its first shot in combat and from then on carried out frequent fire missions, almost daily and at all hours, until the fight for New Georgia ended on August 4, 1943. During that time, no place in the jungle remained consistently secure.[14]

On the night of July 20, a large enemy force attacked the 43rd Division command post at Zanana Beach.[15] Because the enemy was so near to the American headquarters, forward observers called artillery rounds almost on top of themselves. This required a high degree of coordination and control between the FDC, the gun crews, and the forward observers who were "sensing" the rounds as they landed. By using a method of fire adjustment called "sound and fragment," they directed the initial rounds to fall a safe distance forward of their positions, then walked them back by listening for the sound of the impact, bringing the rounds to a zone 100–500 yards from the edge of the bivouac area. The only short round of the entire barrage was one of the initial registration shots.[16]

The 136th laid down a protective curtain of fire around the camp's perimeter all night, breaking up the attack and inflicting heavy casualties on the Japanese.[17] Brig. Gen. Harold R. Barker, 43rd Division commander, who occupied the command post during the attack, was so grateful that the following day he paid a personal visit to the 136th Field Artillery Battalion to thank them for saving his life.[18]

Ground combat has sometimes been called "the meat grinder," and the type of fighting U.S. infantrymen undertook on New Georgia, attacking well-concealed Japanese emplacements, fit the description well. Because the enemy hid their fortifications so expertly, advancing troops generally did not see them until they were literally on top of them.[19] Once aware of their presence, they often withdrew a safe distance and called in artillery or mortar fire to try to destroy them; if that was not possible, they tried to defoliate the ground in front of them.[20] The 43rd Division captured Munda on August 5, ending the most intense phase of the struggle for New Georgia. Yet fighting continued against the remaining Japanese, who offered substantial resistance for nearly two more months.[21]

On New Georgia, American troops witnessed the same ploy Captain Casey had described on Guadalcanal. Realizing that they were overmatched by their American counterparts, the Japanese would wait until the 37th's guns had unleashed a barrage, then attempt to imitate the pattern of fire. One American after-action report noted that frequently throughout the entire campaign, unit commanders would call to complain that their own artillery was hitting their men. Very typically, what these officers deemed as friendly fire was really enemy 90-mm or 75-mm shells.

As they watched their observers adjusting artillery fire, American infantrymen naturally blamed them when they believed that friendly rounds had fallen upon them. In turn, forward observers were generally sensitive to such claims. Sometimes, given the proper circumstances, one could determine the origin of an exploded shell. One American artillery officer on New Georgia described how he was able to confirm that rounds landing inside American lines were Japanese. Immediately after he had called for his initial rounds, the enemy fired upon American positions to his right, convincing the infantrymen there that they had received a short round from their own guns. After all firing had ceased, the forward observer sent an officer to the place where the shell had landed. "I told him to inquire if the smoke from the explosion was black (that's a characteristic of the Jap howitzer shell which I had observed when several of them had sited [sic] our battalion CP one time); the 105 millimeter shell smoke [was] bluish. Well, the witnesses allowed as how the smoke was blackish and furthermore, the officer found the Jap fuse."[22] Unfortunately, not every such incident was so readily verifiable.

One infantry-battalion commander insisted that using artillery and mortars against cleverly concealed Japanese defenses could be very effective while saving numerous American lives, though jungle warfare placed certain limitations upon its use. These included the difficulty in determining the position of friendly lines, the time required to position batteries in such a way as to permit high-angle fire, and the withdrawal of troops.[23] For forward observers, these posed challenges in control and coordination that their training had neglected.

Prior to fighting in the Pacific theater during World War II, U.S. field artillery had very limited experience in jungle warfare. Paragraph twenty-eight of field manual FM 31–20, *Jungle Warfare,* warned that the great weight and bulk of the field guns would hinder their mobility and that the dense growth of the jungles would constrain the explosive burst of projectiles. It added that the density of foliage would greatly reduce the effectiveness of indirect fire because it would limit the visibility of observers whether on the ground or in the air.[24] Lt. Col. Robert C. Gildart rejected these claims. Based on his own experience on New Georgia, he thought that artillery's success in combat there invalidated these claims.[25]

There was some basis of truth in the claims found in FM 31–20, however. Large jungle trees sometimes shielded soldiers from shell fragments. Soft earth and mud were known to help absorb the spread of the explosion of high-angle artillery and mortar shells. In addition, dense jungle growth did hamper observation efforts. Gildart noted that at times observation on the ground was impossible except in portions of the zone of action. Forward observers could not always see the targets on which they adjusted fire. Consequently, field artillery relied on a combination of aerial observation when feasible and ground observers adjusting fire by either sight or sound.[26]

Whether visibility for observation was better from the air or on the ground depended largely upon the topographical characteristics of a particular area. In a rather contradictory statement, the 37th Division Artillery narrative for New Georgia explained that ground observation near Munda airfield was virtually nonexistent. Yet the same paragraph described air observation as particularly good around the airfield (an area cleared of vegetation) while asserting that just to the east, jungle growth precluded effective sighting from the air. In conclusion, the writer cited ground observation as the most satisfactory method of adjusting fire.[27] The conclusion one may derive from this account, then, is that while an observer in an aircraft may have a broader visual perspective and cover much more area in less time than someone on foot, dense jungle growth frequently rendered aerial observation less useful than that done from the ground. Either way, artillery still represented a highly effective weapon in a jungle environment. As noted previously, General Barker felt that he owed his life to the well-placed fires of the 136th Field Artillery Battalion.

Often, two forward observers on the ground, working together but from separate locations, adjusted fire by sound when friendly troops occupied a broad front. Both sent a sensing to the FDC, which then made a comparison and selected a composite to ensure the safety of each man. Observers' greatest difficulty lay in adjusting rounds in the area three hundred yards immediately to their front. All rounds landing in this region seemed to produce the same sound.[28]

Another artillery officer, Lt. Col. Howard F. Haines, in describing the difficulties encountered when adjusting fire by sound on New Georgia, noted that forward observers frequently made their initial sensings by sound. After determining that a round of smoke had landed in a safe location, he would call for a volley of high explosives to try, if possible, to see the bursts. Adjusting fire this way was especially difficult. Sound echoes in the jungle, and sound waves reaching the ear may not arrive from the direction in which they originate. Even night and day, rainy and clear weather make a difference in the way that sound approaches. A shell exploding at treetop level sounds different from one hitting the ground and may, in fact, vary in range over one hundred yards due to the slope of fall. To complicate matters, infantry commanders, unable to see the flash of the burst, typically figured that friendly fire was falling much closer to them it actually was.[29]

Because sound adjustments were comparatively slower than visual sensings, their use also increased the time necessary to mass fires. Colonel Gildart postulated, "Past experience dictated that with this method each battery within a battalion should be adjusted rather than massing fire on the adjustment of one."[30] This was a reasonable precaution, for if the sensings of one or even two observers working together were in error, only one battery rather than an entire battalion might fire short before the mistake was discovered and corrected.

Of course, adjusting artillery fire by sound was not always necessary; visual observation of targets by ground observers was still possible from time to time and was generally very precise. One such incident occurred on July 27, as the 37th Division's artillery was firing a brief mission in support of Lt. Col. Theodore Parker's 2nd Battalion, 145th Infantry. The regimental commander overheard this exchange on the radio as he was awaiting the prelude to the infantry's assault: "'Right twenty-five [yards],' a forward observer with the infantry told his battery executive officer. 'What the hell?' the battery exec countered, 'the effective radius of one round is thirty yards. Why the twenty-five yard shift?' 'Well, I missed one of the little bastards.'"[31]

Aerial observation in the Pacific could be more advantageous than ground observation, but not always. In a section addressing reconnaissance, the 37th Division's artillery narrative explained that aerial observers naturally approached the enemy much quicker than those on the ground, enabling artillerymen to open fire before the enemy could respond. But the advantage and success of spotting the enemy from the air depended heavily upon the terrain. In areas with little cover, like the beaches and the area around Munda airfield, it gave excellent results, but the features of the jungle almost universally obscured targets from aerial observation.[32]

Positioning forward-observation teams with maneuvering troops during World War II led to a blurring of the distinction between the roles played by the two combat arms, infantry and artillery. Lt. Col. Howard F. Haines noted that on New Georgia: "The forward observers and liaison officers stayed with, fought with, and became part of the infantry battalions, going through the campaign with them without relief and then frequently being sent into combat with another battalion when their own was temporarily inactive."[33] The frontline artillerymen also left the line only when they became casualties.

One can readily understand that, regardless of designated role, anyone in the middle of a combat situation will necessarily do what is necessary for self-preservation. For example, when enemy troops ambushed the forward-observation team of 1st Lt. Scott A. McKinnon (of Canton, Ohio) and two men as they adjusted fire in an exposed area, the lieutenant ordered withdrawal. He then covered his men with rifle fire, killing several Japanese soldiers before reaching safety himself.[34]

As forward-observation teams remained with combat infantrymen during World War II, it led to a blurring of rank among these artillerymen. The designated forward observer was a commissioned officer, usually a lieutenant. As the war progressed, however, after so many observers had become casualties, many artillery battalions resorted to using noncommissioned officers as the designated observer. To that point NCOs and enlisted men comprised only the remainder of the team accompanying a lieutenant.

One example of this blurring of rank came on August 5, when Sgt. Robert Harshbarger of Battery C, 135th Field Artillery was leading a forward-observer party. When a Japanese patrol sprung an ambush on his group, Harshbarger used his submachine gun to keep the enemy pinned down. After his comrades escaped, he then withdrew.[35]

Understandable is the human compulsion to help an injured comrade, regardless of one's designated function in a given situation. Amid battle, team personnel frequently did what they could to provide first aid for the wounded. During the assault on Munda airfield, Tech-5 Burt E. Silverthorn of Cleveland, Ohio, served as a member of a forward-observer party. During the night, Japanese artillery shelled their position, wounding all the members of the team except Silverthorn. Leaving cover, he assisted in treating and evacuating the wounded despite being badly shaken by the concussion of a shell that landed six feet away. Then under intermittent enemy fire, he took over for his wounded lieutenant, calling for protective fire that enabled the battalion to hold on during the night.[36] Silverthorn's actions exemplified both the blurring of rank and combat roles.

As mentioned before, the Combat Infantryman Badge and the Combat Medical Badge were two insignia created during World War II to indicate that the wearers had participated in some continuing manner in ground combat. On October 27, 1943, War Department Circular 269 first authorized the CIB, in part, "to foster *esprit de corps* in infantry units."[37] Only infantrymen were eligible to earn and wear the badge.

Originally designated as the Medical Badge, the War Department created the Combat Medical Badge (CMB), on March 1,

Robert Harshbarger, Battery C, at Camp Shelby, Miss., 1941. Photo courtesy of Donald L. Walker.

1945, to "be awarded to the officers, warrant officers, and enlisted men of the Medical Department assigned or attached to the medical detachment of infantry regiments, infantry battalions, and elements thereof designated as infantry." Its evolution stemmed from a requirement to recognize medical aidmen "who shared the same hazards and hardships of ground combat on a daily basis with the infantry."[38] Both the CIB and the CMB quickly became highly coveted awards among those serving in the major theaters of the war, particularly the CIB. It also enhanced the chance of promotion for career soldiers. This, in turn, led to abuses. Gerald Astor called it a "sham" the way in which some officers "earned" theirs: "There were officers who drove into the rear of a combat zone, which could be maybe a mile wide and a mile deep. They'd stay there for a certain length of time, then fill out the papers for a CIB. That gave them the same status as the poor slobs forced to be in the equivalent of the killing zone."[39]

The CIB is not a sham when awarded to those for whom it was intended, and no one who knows anything about the nature of ground combat would ever argue that the combat infantrymen and the combat medics did not deserve these badges of distinction. Unfortunately, the army never created a similar badge for those artillerymen who also "shared the same hazards and hardships of ground combat on a daily basis with the infantry." On New Georgia, forward-observation teams routinely occupied the first fifty to one hundred yards of the front lines, when they were not actually out in front of the combat riflemen.

Typical of almost every island campaign the United States fought in the Pacific theater, New Georgia took longer to complete and resulted in many more casualties than military planners had foreseen. Originally intended to involve only the 43rd Division, it required three army divisions and several marine battalions to take the island. Casualties included 1,094 dead and 3,873 wounded; official estimates of Japanese dead exceed 2,400.[40] Losses for the 37th Division included nineteen officers and 217 enlisted men killed in action or died of wounds and 1,010 wounded. Commanders estimated the division's men killed 1,426 Japanese and took just twenty prisoners. In only twelve days ending on August 5, its field-artillery battalions fired just under 25,000 rounds.[41]

Also typical of the land battles in the Pacific was the way the Japanese generally fought to the death rather than allowing themselves to be captured. The 37th Division's official history does not specify how many, if any, of the twenty prisoners, may have been unconscious or physically incapacitated when captured. In many of the larger contests that followed, twenty prisoners taken would be a large number for a division. The chronicle does tell, however, of several occasions where wounded Japanese soldiers played dead, then opened fire on American soldiers as they passed by, knowing their own deaths were inevitable.

After the war General Beightler summed up the New Georgia campaign, calling it a "shoe-string affair" compared with the subsequent campaigns in which his division fought. The Japanese impressed Beightler with "their ability to organize strong defensive positions and their willingness to fight to the last man. But he also thought that his men were as good fighters as the Japanese and, when backed with the U.S. Army's superiority in artillery, were unbeatable."[42]

After the war Beightler expressed the respect he had gained for his own artillery in the summer of 1943 and the benefits derived from relying on it heavily, calling New Georgia "the testing and proving ground" for the division's guns. He concluded that his superiority in artillery tipped the scales in favor of his division in battle, and (as noted earlier) strongly believed that the number of casualties his men incurred was inversely proportional to the amount of artillery shells fired.[43]

As a veteran of World War I, General Beightler knew something about the human cost of winning battles. Inculcated with General Pershing's doctrine of open warfare that emphasized the primacy of the rifle, Beightler was realistic enough to understand that infantry alone could not overcome well-concealed machine-gun emplacements. He was also aware that the soldiers of the AEF succeeded not from any decided tactical advantage, but as James W. Rainey notes, "by smothering German machine guns with American flesh."[44] What Beightler saw regarding the level of success in the execution of combined-arms tactics on New Georgia encouraged him to rely more heavily upon his artillery to reduce his division's infantry casualties in subsequent campaigns. His comments allude not to the achievements of artillery alone, but to the newfound success field artillery experienced fighting in combination with infantry during World War II, success on a level not achievable only twenty-five years earlier.

For the 37th Division, New Georgia represented its initial opportunity to execute combined-arms warfare in action. For the Buckeye Division's artillerymen, it was their first chance to begin ironing out problems of command, control, and communications in the practice of such tactics. Of the three, problems with command were minimal as the new decentralized system greatly facilitated the massing of fire.

The division's forward observers experienced problems immediately with control, however, both in observing and adjusting fires. From the realization that the teams must be up front with the infantry came the eventual blurring of distinction between combat arms that, for the 37th, began on New Georgia. In a jungle environment where, at times, it was impossible for a man to see more than twenty feet ahead, the observers developed the ability to use sound rather than sight to sense the fall of rounds. Finally, with regard to control, they had to deal with the distinction between ground and aerial observation and the advantages and disadvantages inherent in both. But this was not a problem unique to either the 37th Division or the Pacific theater.

Visual sighting of targets on the South Pacific islands remained problematic during World War II, much as the use of observed fire in general had been throughout the Great War, but for a different reason. Now the nature of the terrain, not the placement of the observer, was the reason. Bougainville, where the 37th Division would next fight, was no different in that respect than New Georgia. The battalion intelligence officer for the 129th Infantry recalled that during the Bougainville campaign, he often flew as an observer in a small plane and that other than the natural beauty of the island, "there was little to see under the dense cover provided by the trees."[45] Such was the typical nature of the terrain on the islands. While an observer in an airplane had a much wider view than a man on foot, that meant little in zones concealed by the thick tropical vegetation.

The problem of control through observation was a perpetual problem on New Georgia, while problems with communications, although troublesome, were not quite as perplexing. Although the U.S. Army was making extensive use of portable field radios by this time, forward observers on New Georgia depended on telephones more than they did radios. At times, however, phone lines were broken, typically because friendly, not enemy, troops had inadvertently cut them. As crews widened or rerouted the roads in areas away from the fighting, they felled trees; when the trees came down, the wire came down with them. Repair crews were almost always at work, two men repairing while three guarded against possible snipers. All this additional support activity meant that the number of personnel required in the jungle was far in excess of that authorized in the newly created Tables of Organization and Equipment, drawing artillery's pool of manpower even thinner.[46] Yet U.S. artillery and infantry forces on New Georgia relied heavily upon telephones because, as Colonel Gildart explained: "We only have one radio light enough for an observer to carry and yet still powerful enough to communicate through the jungle to the FDC. This set is the SCR-511 (used in conjunction with an SCR-284 as a base set); and it is fragile and susceptible to dampness."[47]

Not available during World War I, portable field radios held great promise. But in the field, they did not always function properly or consistently and were difficult to maintain under less-than-ideal conditions. But the single most significant improvement in forward observation was achieved simply by placing those responsible for observing and adjusting artillery fire at the front. Despite difficulties in visual observation and maintaining communications, forward observers were no longer somewhere between the batteries and the front lines but rather in the foremost positions possible—in some cases at Munda, actually in advance of the lines.[48] Nor did the noise and confusion of battle prohibit them from effectively adjusting fire during combat as some had supposed might happen.

The division's use of field artillery on New Georgia seemed to indicate that all the limitations that had prevented field artillery from providing close support of infantry during World War I had either been eliminated or would soon be worked out. From their new position within the front lines, forward observers did more than just call for and adjust fire to contribute to winning the war: they took part in the fighting.

For *Field Artillery Journal*, General Barker listed what he considered to be the most outstanding features of the New Georgia campaign. Noting that the tactical employment of artillery on the island conformed closely to the procedures spelled out in existing manuals, he wrote that "one of the most gratifying results of the New Georgia campaign from the artilleryman's viewpoint was the confidence and enthusiasm displayed by our infantry for their artillery."[49] At times unit commanders wanted artillery fire placed as close as fifty yards from their front lines. The general also noted the importance of the forward observers to use sound adjustment, adding that at times it was the only way results could be obtained and that the men became very proficient at it.[50] In his final analysis, Barker praised "the superior work of the personnel employed in maintaining contact with the front line infantry units such as liaison detachments, FOs, and communications personnel," commenting that "at least sixty percent of all artillery officers were with the infantry on the mainland at all times." Finally, he lauded the ability of American artillery to mass fires at battalion, division, and corps levels.[51]

World War II witnessed the development of the close coordination between infantry and artillery that was lacking during World War I. For example, on New Georgia, American artillery fired rolling barrages like those employed in the previous war, beginning close to the infantry and moving ahead 500 yards in increments of 50–100 yards. When the troops followed closely, they typically incurred few if any casualties. But when they delayed their departure from the front lines by an hour or more, they took heavy casualties from Japanese automatic weapons and made little, if any advance.[52]

In summary, what New Georgia proved was that American infantrymen on the attack could rely on observed-fire support to accommodate each emerging situation they encountered. Although fire support during World War I could be effective, it lacked the responsiveness to provide spontaneous help and to quickly mass fire for that matter. The gun crews, FDC, and field radios all played important parts, but forward observers proved necessary to make it work. As one army publication observed, the efforts of forward observers on New Georgia was "outstanding."[53]

The experience of the 37th Division on New Georgia also demonstrated the inequality between Japanese and American artillery. During the battle for Bougainville, this imbalance would become even more pronounced. Japanese infantry doctrine and tactics would put their attacking riflemen at a decided disadvantage against the Americans' ability to mass fires effectively.

CHAPTER 5 | # Massed Fires and Mass Slaughter

Bougainville

The battle for New Georgia gave the forward observers of the 37th Division their first opportunity to apply their training to battlefield conditions. For the most part, it served them well. The major exception, however, was the frequent need to adjust fire by sound rather than by sight. Their first taste of combat also provided them with a close look at an extremely cunning enemy. The Japanese trick of firing a few shells into American positions between salvos made forward observers even more cautious in carrying out their work. The fall of New Georgia provided additional airfields from which to launch strikes against the Japanese, but enemy aircraft still posed a threat from bases on Bougainville, Buka, and Rabaul.[1]

Although not one of the well-remembered battles of the Pacific War, for the men who fought it, Bougainville became an ordeal and an experience they most likely never forgot. U.S. military planners considered its capture in 1943 absolutely essential in preparation for the invasion of New Britain and the reduction of Rabaul, Japan's largest air and sea base in the South Pacific.[2] Bougainville provided a much greater test of the 37th Division's ability to combine infantry and artillery tactics in the difficult South Pacific jungles. This campaign thoroughly tested control and communications. Its most distinguishing aspect—massed fires and mass slaughter—came with both combat arms operating to their fullest potential in defensive tactics. The carnage that resulted demonstrated the disparate nature

BUKA
ISLAND

Buka

Bonis

Soraken

Sapaai

Kuraio Mission

Laruma
River

EMPRESS
AUGUSTA
BAY

SOLOMON
SEA

Cape Torokina

Mosigetta

Hari
River

Silibai
River

Buin

Kahili

Mt. Balbi

Numa Numa

Arty
Hill

Mt. Bagana

Kieta

SOUTH
PACIFIC
OCEAN

SHORTLAND
ISLANDS

0   5 10   20   30   40
Scale of Miles

Bougainville Island. From Harry A. Gailey, *Bougainville, 1943–1945: The For-gotten Campaign* (Lexington: University of Kentucky Press, 1991).

of the ground war in the Pacific and the way in which a flawed Japanese infantry doctrine played into the hands of American military strength.

The northernmost and the largest of the Solomon Islands, Bougainville is 125 miles long and 30–48 miles wide.[3] Its topography favors defensive action, with high volcanic mountain ranges, dense rainforests, thick jungles, swamps, and a few broad plains. But the island is about five times larger than New Georgia, with more open area. Although dense jungle growth would again obscure targets, the open spaces would increase opportunities for forward and aerial observers to adjust fire by sight rather than sound.

The American invasion of Bougainville began when elements of the 21st Marine Regiment landed at Torokina on November 6. The 148th Infantry landed on November 8 and was assigned to the 3rd Marine Division. The 129th and 145th Infantry followed a few days later. Most of the 37th Division's artillery had landed near Torokina by midmonth, including the 136th Field Artillery on November 17 and the 135th two days later.[4]

November proved largely uneventful for the Buckeye Division except for its artillery units. By the twenty-third, marine and army reconnaissance had designated an area approximately eight hundred yards square for artillery bombardment to precede an assault by two battalions of the 3rd Marines. The next morning seven howitzer battalions—four marine and three from the 37th Division—opened fire at the appointed time with a concentrated barrage. A half hour later a Japanese battery began directing accurate fire on the assembly positions of the marines. Soon the enemy artillerymen found themselves overmatched.[5]

A forward-observer party on Cibek Ridge spotted the enemy position and called for counterbattery fire from 155-mm howitzers of the 136th Field Artillery Battalion, which quickly destroyed the Japanese guns.[6] American artillerymen fired a total of sixty tons of high explosives upon enemy emplacements. The 135th Field Artillery Battalion's history notes that the division's artillery fire on that date killed an estimated 1,021 enemy soldiers.[7] Japanese prisoners later indicated that those not physically injured by the intensity of the barrage had left the line requiring treatment for extreme neurosis.

In the Pacific theater, army forward observers, at times, directed fire for the navy and Marine Corps.[8] Today the field artillery is known as the army's integrator of joint fires and effects, with the ability to coordinate the firepower of other branches with its own.[9] In 1943 the practice of using joint fires was still new.

On the morning of November 29, Maj. Richard Fagan's 1st Marine Parachute Battalion, accompanied by Company M, 3rd Raider Battalion and a forward-observer team from the 12th Marines, carried out a raid against Japanese installations east of the Saua River. The party came ashore "virtually in the middle of a Japanese supply dump."[10] The immediate enemy response left the marines pinned down close to the beach for hours. A forward observer from the 136th Field Artillery Battalion then adjusted artillery fire on the Japanese positions, enabling the marines to withdraw to safety that night.[11] On November 30 Capt. Milton Bagby of the 136th adjusted the fires of two naval destroyers as they shelled Japanese positions along the coast from the Saua River east.[12]

By mid-December 1943 the marines, with support from the army, had carved out an area roughly four-and-half miles by six miles and had constructed three airstrips within the perimeter. Near the end of the month, the Americal Division arrived at about the same time the 3rd Marine Division departed. As he was leaving the island, the marine artillery commander expressed his gratitude to the U.S. Army for its cooperation and praised the efficient conduct of artillery on Bougainville, calling it "an example of exceptionally smooth operation between two units of different Federal forces," adding, "[it is] also evidence of a spirit of comradeship and cooperation worthy of the best traditions of the United States military forces."[13] The 3rd Marine Division combat report for the Bougainville operation also lavishly praised the artillery support furnished by the 12th Marines and the 37th Division, describing their accurate fire as "the dominant factor in driving the Japanese forces out of the Torokina area." The report credited at least half of the enemy's casualties to artillery fire.[14]

The Americans met less enemy resistance than they might have when they arrived on Bougainville in November because Japanese general Hyakutake thought the initial landings represented a feint and that the bigger landing would come later. Hence he had declined to shift the bulk of his troops to the area around Empress Augusta Bay. By February 1944 the commander of the Eighth Army Area (headquartered on Rabaul), Gen. Hitoshi Imamura, directed

Hyakutake to prepare for an offensive, codenamed TA, to break through the American perimeter and capture the three airfields on the southwestern side of the island.

Japanese intelligence grossly underestimated U.S. troop strength, figuring that the 15,000 men Hyakutake would have available, supported by the largest concentration of Japanese artillery ever assembled in the Solomons, could succeed. Instead, by the beginning of the new year, Maj. Gen. Oscar Griswold, commander of all U.S. forces on Bougainville, had more than 50,000 troops to hold the ground gained, a figure more than doubling Japanese estimates.[15]

Every manner of Allied intelligence, from coast watchers to long- and short-range ground patrols, prisoners, and even a few deserters, indicated that the Japanese were marshaling their forces for a large-scale offensive. By mid-February patrols and brief fire fights indicated that the enemy was concentrating troops around the top third of the arc formed by American forces defending Cape Torokina. Papers taken from enemy corpses gave Gen. Griswold the exact details of the Japanese plan of attack, and other valuable information including the general location of enemy artillery units.[16]

From these captured documents, American intelligence learned that the Japanese planned to attack on March 6, later delayed until the eighth. With a huge advantage in men, the only American disadvantage was position. The Japanese held most of the high ground in front of the American zone, enabling them to observe activity along the perimeter. To execute TA, General Hyakutake called up the 6th Division.[17] The 6th had seen long service in the war in China and had taken part in the infamous Rape of Nanking.[18]

At about the same time, evidence from patrols indicated increased Japanese activity in front of the 37th's lines near the Tsinamatu River. Division headquarters decided to contest the area and sent two companies from the 145th Infantry to reinforce those already in place beyond the perimeter. A force of about five hundred men advanced about a half mile before digging in and requesting artillery fire for the morning of February 16 prior to making their attack on Japanese positions. Harry Gailey has noted that some of the rounds fell short, killing three members of the 145th and wounding twenty.[19]

This may have been another case where the Japanese threw in a few rounds to make the Americans think they were being hit by their own fire. Forward observers during World War II carried a heavy responsibility. Unquestionably, some were responsible for friendly fire casualties through their errors. Today those directing artillery use computerized range finders to adjust fire, but in 1944 they did it by eye or ear. In this case those who were there have branded this incident an artillery failure. Nonetheless, the observer's instructions alone do not necessarily determine where a round lands. Bags of gunpowder propel an artillery shell. These are numbered and so many bags of a particular lot number are used according to the range to the target. If a factory worker in an ammunition plant fails to put the correct amount of gunpowder in a bag or mislabels it, when fired the shell will not land where intended.[20]

The famous war correspondent Ernie Pyle once witnessed a near calamity involving powder charges during a visit to an artillery battery as it was about to fire a mission. As a member of a gun section picked up a powder bag, he noticed that it was only half-filled with gunpowder. The artilleryman said to Pyle, "If we'd shot that little one, the shell would have landed on the battery just ahead."[21] In the timeless dimension of combat, fatigue could become a problem when gun crews worked around the clock. Describing his concerns for the ammunition his unit used on Bougainville, Capt. Robert F. Cocklin wrote: "Our biggest difficulty is keeping the lot numbers straight. Quite often the difference in lot numbers makes a considerable difference in the range of rounds fired, so too much emphasis on this angle is impossible."[22] If a tired member of a gun crew picked up the wrong lot numbers during a fire mission, the resulting shot would not land where the forward observer and FDC had calculated that it would, and no one might ever know the real reason. But how many times this actually happened during the war is speculation.

Hill 700, on the right and held by the 145th Infantry, represented the highest ground secured by the Americans.[23] The Japanese plan of attack called for two units to stage a series of attacks, beginning on March 8, along the perimeter and to advance to Piva airfield, followed by a third force attacking on March 11. This was an overly ambitious plan because the Japanese were severely understrength in men and artillery and had no air support whatsoever.[24]

The attack began on the eighth, with the bulk of its force being directed at the 37th Division. Some of the bloodiest fighting on

Bougainville took place over the next two weeks as the Americans and Japanese engaged in a series of violent, see-saw engagements to hold several lines of pillboxes situated on the hilltops. Hill 700 became the chief objective of the Japanese 23rd Infantry. Because of its steep slope, Beightler had not anticipated the enemy would concentrate their attack there, but apparently they wanted the high ground.[25] In the first five days of battle alone, the 37th Division lost five officers and seventy-three enlisted men killed. Artillery fired in defense of Hill 700 amounted to 20,802 105-mm rounds, 10,000 75-mm rounds, 13,000 81-mm rounds, and 811 4.2-inch shells.[26]

An interesting phenomenon occurred the morning of the opening attack: both sides used direct-fire artillery. The Japanese probably had to, and the Americans did it because it was to their advantage to do so. At 6:15 on the morning of the attack, Japanese 150- and 75-mm guns began shelling American positions along the Saua River and the adjoining hills. Because the Japanese were using direct fire, American forward observers were able to spot their gun flashes and adjust effective counterbattery fire. While the 6th Field Artillery Battalion and the 129th Infantry Cannon Company were able to shoot directly at the enemy, the 135th, 140th, and 136th Field Artillery Battalions fired indirectly, their barrages adjusted by a forward observer. Effective American counterbattery fire quickly destroyed or silenced many enemy gun positions.[27]

Japanese artillery on the morning of March 8 caused few American casualties. As the enemy began to shell the Piva airstrip, Lt. William D. Jennings of Toledo, Ohio, forward observer with the 135th Field Artillery, immediately spotted the gun flashes, and after he had made his adjustments, the 135th hit the Japanese with a heavy counterbattery barrage.[28] Frankel noted how afterward one prisoner sadly remarked: "Each time we fire one round, you send back a hundred in return. No good."[29]

That evening four battalions of 37th Division's artillery and two of the Americal Division prepared to fire on Japanese troops assembling behind Hills 1111 and 1000. At the designated time, an intense concentration of artillery fire lasting for two hours smothered this zone of activity. A Japanese prisoner later reported that this barrage had been so intense that he and his comrades had moved as close as they could get to the American lines to avoid annihilation. Anticipating this Japanese reaction and to preclude endangering friendly troops, forward observers had moved the fire ever closer to the 37th Division's front lines. On this day alone, the 136th Field Artillery fired 1,239 rounds, nearly ten times the average daily number.[30]

The forward observer's job was hazardous not only because of his proximity to an enemy trying to kill him but also because he had to position himself close enough to see his own artillery's shell bursts. When necessary to stop an enemy attack, most would not hesitate to call for fire virtually on top of their own positions. Despite suffering heavy losses overnight, the Japanese continued their attack on the morning of the ninth. At 7:30 A.M., aware that the enemy was advancing toward his position, Lt. Robert P. McClendon, forward observer with Battery F, 135th Field Artillery, alerted the battalion and instructed them to "pour it on as close to me as you can get it."[31] Two other battalions joined the 135th in this mission. All of the 135th Field Artillery Battalion's forward-observation teams were working with separate infantry units in the field that day.[32]

Early on March 13, the Japanese mounted an attack against the lines of the 145th Infantry, directing heavy machine-gun and rifle fire against them while charging Observation Post 5, manned by artillerymen of the 135th Field Artillery. "With the officer adjusting artillery fashion, one enlisted man threw grenades (all the while exposing themselves) while the other burned out several carbines felling the swarming enemy, one of whom almost reached [the] OP dugout."[33] Although they did this to save their lives, the incident again demonstrates the blurring of the distinction between infantry and field artillery during World War II.

The beginning of the end for the Japanese on Bougainville came on March 17, when they began their final series of attacks. Following rapid penetration into the American lines, the fighting again became a back-and-forth struggle for a few days. Then came a lull while the Japanese brought in reinforcements. During the evening of March 23, they began to advance in what became their last serious effort to infiltrate the American perimeter.

Early on the morning of March 24, the Japanese made their strongest attack and deepest penetration. After having cut off their advance with an artillery barrage, a combined U.S. infantry-tank counterattack drove the enemy back and restored the lines. The Japanese were now finished but did not know it. American intelligence had obtained vital information the previous evening regarding details of the attack, including where and when it would begin. General Beightler then visited General Griswold and, after relaying the news of the impending attack, asked for an extraordinarily large supply of ammunition and permission to receive priority for all calls on artillery within the perimeter. Griswold granted both requests.[34]

At 10:45 that morning, all the artillery battalions of the 37th plus three of the Americal Division opened fire on the Japanese front and rear areas.[35] Stanley Frankel has observed that what followed was "the heaviest artillery concentration yet seen in the Pacific war, completely destroying all Japanese hopes of regaining control of Bougainville."[36] At one point American artillery rained more than 4,000 shells upon the Japanese forward-assembly area within a single fifteen-minute period and a total of 14,882 rounds altogether.[37] At 6:00 P.M. a counterpreparation barrage was fired and another at 8:16. Finally, at 5:30 the next morning, the last of the counterpreparation barrages were fired. The intensity of this bombardment was the result of information obtained from an enemy prisoner who indicated that the Japanese had massed the last of their surviving forces in a desperate effort to break through the American lines.

Undoubtedly, this tremendous volume of explosives completely frustrated the Japanese offensive and dashed their hopes for holding out on Bougainville.[38] Frankel noted that the gigantic bombardment put an end to this particular phase of the fighting because in the days that followed, all Japanese offensive actions were weak and disorganized. Surviving enemy soldiers pulled from pillboxes exhibited signs of shock. Where the artillery concentration had fallen, the effects of carnage and destruction appeared particularly gruesome. On March 27 what was left of the 45th Infantry on Bougainville withdrew over the mountains.[39]

Overconfident Japanese commanders had not only underestimated American troop strength but also ignored the potential destructive threat posed by American artillery. When a military unit as large as an infantry battalion masses along a narrow front less than one hundred yards wide, by its sheer numbers it becomes difficult to stop even with the use of automatic weapons such as rifles and machine guns, though not so hard with well-directed field artillery. It mattered little whether the troops were extremely brave or simply overzealous because the Japanese could not hold what they had gained in March due to superior American firepower. (Nearly sixty years later the U.S. Army would use even more powerful combined-arms tactics on the offensive in Iraq.) This massive display demonstrated on Bougainville in 1944 certainly represented an early demonstration of "shock and awe." From March 8 through the end of the month, the 135th Field Artillery Battalion fired 25,464 rounds in support of the 129th and 145th Infantry Regiments and also into the Americal Division's sector. All observers and gun crews served with great efficiency.[40]

Field artillery wire crews and linemen during World War II never received the credit due to them for performing a particularly dangerous job. They often accompanied the forward-observation party. Linemen were not involved directly with adjusting fire. Instead, they had the extremely important job of keeping the forward observer's telephone lines open by splicing breaks as they occurred or became known. Because anyone observing them could tell what they were doing, linemen also became important targets for the enemy.

On March 10 two lineman from Battery C, 135th Field Artillery, Pvt. Russell D. "Skeeter" Wright of Alliance and Pvt. Harold F. Morrow of Sebring, Ohio, were accompanying a forward-observation team as it came under heavy enemy fire. The two men strung wire from a covered dugout to a new observation post on the side of a hill, all the while under Japanese fire. They had just completed their hazardous work when a mortar shell landed nearby, killing Morrow instantly and severely wounding Wright, who died three hours later. Later their comrades found two destroyed enemy mortars and numerous bodies in the same area of the artillery concentration the two had helped bring to bear. Both soldiers received the Bronze Star posthumously for their unselfish efforts.[41]

Although a commissioned officer was designated as the forward observer, NCOs and enlisted men frequently did the job. On March 10 Staff Sgt. Frank Edwards, a member of Battery C and also from Alliance, Ohio, was helping the forward observer adjust fire from an exposed position when a Japanese mortar shell exploded amid the group, wounding the officer and killing four men. Although dazed by the blast, Edwards helped evacuate the wounded and reorganize the remainder of the party. After reestablishing communications, he directed artillery fire until another party relieved the survivors of the original group.[42] The medals awarded to 37th

Division artillerymen include many citations for enlisted men performing the duties of the designated forward observer.

During the war, the U.S. Field Artillery's Officer Candidate School at Fort Sill placed special emphasis upon teaching its students leadership skills. This paid handsome dividends when these young officers later entered combat. The actions of Lt. John Robohm of Minneapolis, while serving as a forward observer on Bougainville provide a good example. On March 20 the Japanese ambushed the infantry patrol Robohm was accompanying, killing the patrol leader and four men and cutting their wire communications. The young lieutenant took command of the patrol, ordering its withdrawal and bringing a wounded infantryman with him, and covered the men with rifle fire as they filtered back to safety.[43]

General Beightler later wrote regarding the action on Bougainville, "[n]ever before had more frightful or bloody fighting taken place in the Pacific," estimating the number of Japanese killed as being more than 10,000 men.[44] He had a genuine concern for the welfare of his men, for he made every effort to limit the number of casualties while carrying out the division's assigned mission. Referring to the 37th's entire combat experience, including New Georgia, Bougainville, and Luzon, Beightler later wrote: "We refined our policy of letting machines fight for us to the maximum. For instance, we shot up more than 450,000 rounds of artillery. The dividends that helped pay is exemplified in the fact that we killed Japs at a ratio of thirty-three to every American soldier lost."[45]

Although the Japanese maintained a presence on Bougainville until the end of the war, by the summer of 1944, they and the Americans coexisted under an unspoken truce of sorts. Despite the intense hatred each side had developed for the other, not all contact between them was violent. Close to Piva airstrip, the Americans built a baseball field where they played games on a regular basis after the fighting had died down. One day someone noticed a lone Japanese soldier, a forward observer of sorts, sitting along the edge of the jungle beyond right field watching the game. Soon he became a regular fan, rooting for the 37th Division's teams and somehow indicating his approval for their hits and runs. If he liked baseball, they figured he could not be all bad—no one informed any gung-ho officers of his presence either.[46]

On Bougainville the 37th Division had the opportunity to execute combined-arms tactics beyond the scope of infantry-artillery coordination. It conducted joint fires with the navy and the Marine Corps with equally successful results. The artillery experience there was also a model for massing firepower to break the strength of human-wave attacks. As such, it provided lessons for the use of combined arms in the defense.

The campaign also gave the 37th extensive experience resolving the problems encountered in carrying out close coordination of infantry-artillery tactics. These included the problems of control associated with short rounds, dealing with the enemy's deceptive imitation of U.S. barrage patterns to emulate short rounds, and the new experience of controlling defensive fires. In the latter instance, forward observers typically had less of a problem seeing their targets but had to bring fire much closer to friendly lines than the normal tolerance. Frankel described as "a threadbare Jap trick moving close enough to our lines to get within the umbrella of safety."[47] Later during the Vietnam War, the North Vietnamese and Vietcong would execute the same tactic, which they described as "grabbing the Americans by the belt."[48]

Although Bougainville gave the 37th Division its most frightening experience using artillery in defense of its own positions, the weakness of Japanese tactical doctrine played into American hands with their newfound strength in infantry-artillery coordination. Massed attacks left Japanese infantrymen with no protection from American firepower, demonstrating the mismatched nature of the Pacific War at the doctrinal level. The Japanese faith in the courage and ability of their soldiers was inadequate in the face of well-orchestrated artillery.[49]

Forward observers in the defensive posture found themselves even further integrated into the infantry combat arm. Their responsibility to perform a technical function by providing artillery support for the men around them and the exigencies of individual combat situations that often required fighting as riflemen to save their own lives blurred their original role as artillerymen. The Field Artillery School's emphasis on training newly commissioned officers who would eventually serve as forward observers to be leaders demonstrated the adage that in order to lead effectively, one must be able to lead at the front. As mentioned earlier, during the war in Vietnam, good forward observers typically became the infantry company commander's right-hand man, a role already being established during World War II.[50]

The artillery experience on Bougainville provided an important early example of the conduct of joint fires between different service branches in its infancy. The 3rd Marine Division's combat report paid tribute the effectiveness of these joint fires, describing this coordination as "one of the highlights of the Bougainville campaign."[51] Army infantry captain John C. Guenther had similar praise, observing that the employment of artillery was essential to the successful coordination that represented "the critical point in our superiority over the Japanese." He added that on Bougainville, "the operations of the 37th Division Artillery and of the Artillery Group Headquarters established and commanded by Brigadier General Leo Kreber are an important chapter in the textbook of jungle warfare evolved by the American Army in battle."[52]

Finally, the accounts given by individual Japanese soldiers who survived the deadly barrages provide perhaps the best evidence of the early display of shock and awe wrought by 37th Division's artillery. Most testify to the accuracy and effectiveness of American counterbattery fire. All speak of the extreme level of casualties their units incurred. The capacity to mass fires in coordination with maneuvering infantry enabled U.S. military forces to defeat the Japanese on Bougainville. Without forward observers on the ground with the troops and in the air over the island, this deadly and effective coordination of artillery fire could not have been achieved.[53]

| # Collateral Damage in the Philippines

Luzon and Manila

The Republic of the Philippines is an island nation located in the Malay Archipelago. It consists of more than 7,000 islands covering an area in excess of 100,000 square miles. The islands are divided into three groups: Luzon, Visayan, and Mindanao. About two-thirds of the Filipino population lives in the Luzon group.

Except during the fight for Manila and last weeks of the war, the 37th Division's conduct of artillery fire on Luzon was much the same as it had been on Bougainville and New Georgia, though with some exceptions. With far more inhabitants than either of the others, Luzon had more land under cultivation, providing aerial observers with increased visibility and expanded opportunities to find and adjust fire on targets. Also, with complete air superiority, tactical air support could be devoted to supplementing artillery.

The Japanese had stockpiled artillery ammunition earlier on Luzon, and when the invasion did come, their guns were well supplied and used more extensively. As a result Japanese artillery inflicted heavier casualties on U.S. troops than the 37th had previously experienced. Yet the inequality in tactical doctrine that existed between the opposing forces on New Georgia and Bougainville was equally obvious in the Philippines. Japanese coordination of artillery with its infantry and other commands was glaringly deficient. The Imperial Army practiced poor command, weak control, and inadequate communications. Maj. Archibald Rogers, who fought

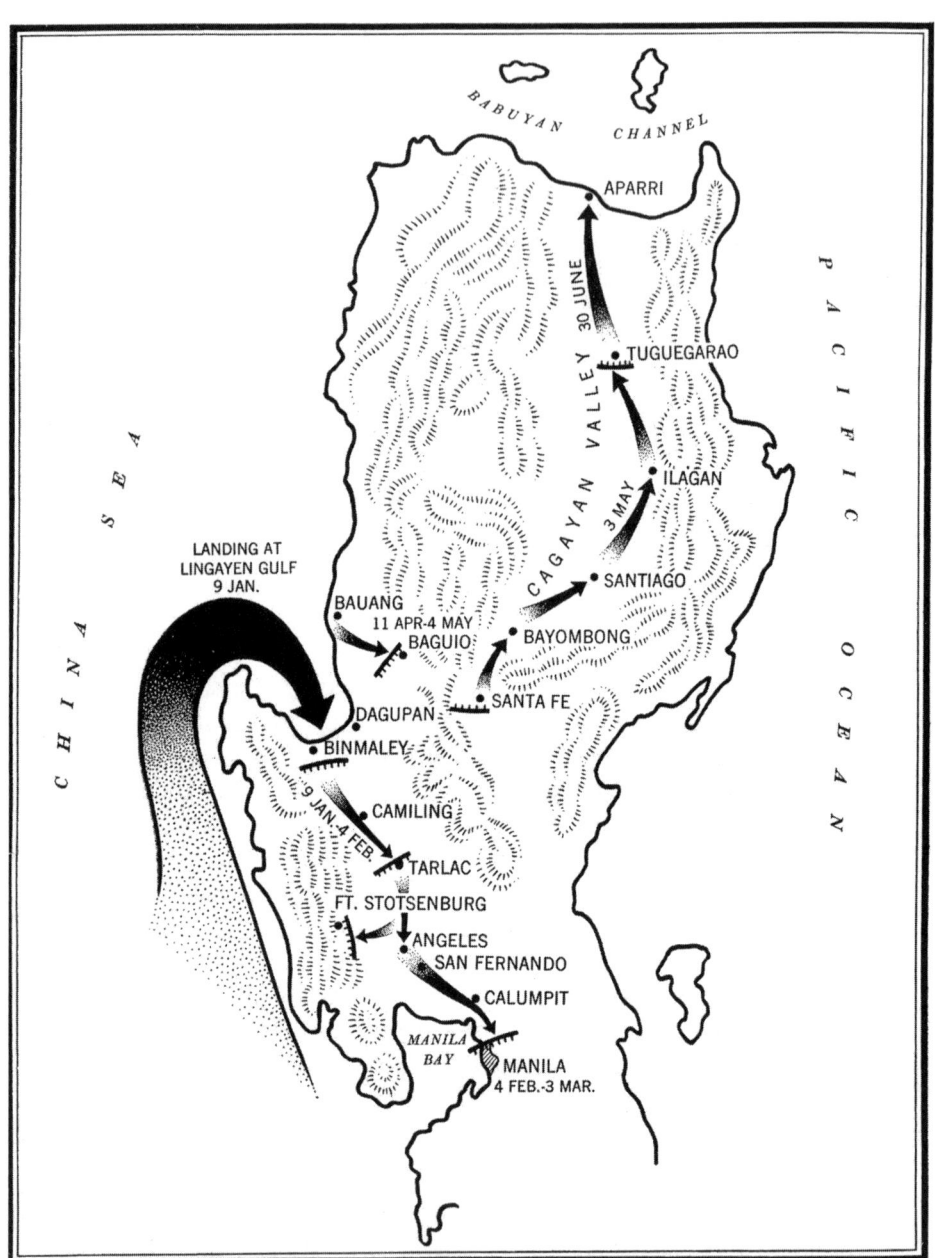

The Luzon Campaign. From Stanely A. Frankel, *The 37th Infantry Division in World War II* (Washington, D.C.: Infantry Journal Press, 1948).

on Luzon, attributed this to "poor liaison, lack of communication, and inability to depart from a prearranged plan."[1] Throughout the Luzon campaign, Japanese artillery never matched the 37th's ability to coordinate its fire support with maneuvering infantry or to engage in effective counterbattery missions.

The real anomaly on Luzon was the fight for Manila, the only sizeable city in which Americans engaged in ground combat during the entire Pacific War.[2] The 37th Division carried the heaviest casualty burden among U.S. forces, yet an estimated 100,000 civilians died in the struggle to liberate the Filipino population and American prisoners from the Japanese on Luzon. Most of the fighting in Manila took place in close proximity to the civilian population due to deliberate enemy positioning. The 37th continued to rely heavily upon field artillery to support infantry assaults. In the city the division made substantial use of direct fire to destroy Japanese strongholds.

Based on the premise that the Allies might have to invade Japan to end the war in the Pacific, the Joint Chiefs of Staff had designed a plan by February 1944, designating Formosa as a necessary stepping stone to the China coast; hence the logical place to establish a base of operations in preparation for invading Japan. Proponents of this plan saw no need to invade the Philippines.[3] By the end of July, however, the Joint Chiefs had agreed that American forces would strike into the south or central Philippines before advancing to either Formosa or Luzon. But if commanders were to choose Formosa, they would have left Luzon, the most heavily populated region of the Philippines, in Japanese hands. As it happened, in mid-August the Japanese began overrunning the last air bases in China from which the U.S. Fourteenth Air Force could effectively support invasions of either Luzon or Formosa. This changed the situation, invalidating the main reasons for invading the latter to use as a stepping stone to China. On October 3 the Joint Chiefs authorized Gen. Douglas MacArthur to begin the invasion of Luzon a few days before Christmas.[4] Yet the preliminary invasion of the Philippines at Leyte on October 20 took longer than expected, and the new invasion date became January 9, 1945.[5]

The 37th Division's landing at Lingayen Gulf took place as scheduled, almost unopposed.[6] During its first week on Luzon, the division encountered only light Japanese resistance.[7] At the same time, MacArthur was so encouraged that he concluded that Gen. Walter Krueger's Sixth Army would reach Clark Air Field within a few days and possibly enter Manila by January 26.[8]

On January 27 XIV Corps issued orders for a coordinated attack by the 37th and 40th Divisions to secure Clark Field, Fort Stotsenburg, and the adjacent high ground known as "Top of the World." The Americans moved out at 7:00 A.M. the next day. The 145th Infantry had reached its objective by 1:00 P.M., but the advance of the 129th was slowed by heavy fire from the eastern edge of Fort Stotsenburg. The 135th and 136th Field Artillery Battalions fired direct-support missions against enemy machine guns, antiaircraft guns, and soldiers, using aerial observers to adjust fires. During this time, the Japanese retained the high ground and, with superior observation, were able to direct counterbattery fire against all engaged U.S. batteries. This was one of the few instances where the Japanese had much success of this kind against the 37th Division, though the enemy guns were eliminated as soon as they were located.[9]

In the early stages of the battle, the U.S. Army relied heavily upon aerial observation for finding targets and adjusting artillery fire. Due to the lack of ground opposition, the planes were able to circle right over the target, making artillery fire very simple and extremely accurate. Because of the open terrain of the central plains between Lingayen Gulf and Manila, aerial observation often worked well in this area. But the soldiers of the Imperial Army were masters of camouflage and concealment, and because of this, target acquisition and fire control always depended heavily on ground observers. Aerial spotters had difficulty seeing well-hidden targets from a plane, but an observer on the ground with a map could easily make accurate adjustments.[10]

Although aerial observers typically had a better vantage point for spotting enemy activity, at times it could be quite difficult to spot targets from the air. The Japanese constructed heavy earthen-and-timber pillboxes at crucial points such as river crossings and road junctions along the American route of advance.[11] In hilly areas they dug their foxholes on narrow ridgelines and used caves and tunnels on the crests or reverse slopes to build field-gun emplacements. Many of their artillery positions were in tunnels with entrances on the reverse slope and only a small aperture for the barrel on the forward slope. Unless American airborne observers could detect the

John Wendell Josh, Battery C, at Camp Shelby, Miss., 1941. Photo courtesy of Donald L. Walker.

muzzle flash, it was difficult for them to sense the location of the Japanese pieces.[12] Observers on the ground, however, could occasionally hear the sound of the enemy guns.

During the battle for Clark Field on January 29, Japanese artillery fire repeatedly cut American telephone lines.[13] During the action, Cpl. John Wendell Josh, a member of Battery C, 135th Field Artillery and a radio operator with a forward-observer party, voluntarily crossed open terrain in full view of the enemy to reach a position from which his lieutenant could adjust an artillery strike. Although his bulky equipment drew enemy fire as he ran, he reached the observation point safely, and the team was able to direct a devastatingly effective bombardment on Japanese targets.[14]

Josh was a brave soldier and one of many who contracted malaria while serving in the Pacific, suffering recurring bouts for many years thereafter. Les Boren, also a member of Battery C, noted that a few years after the war, Josh was in a movie theater in his hometown of Alliance, Ohio, when he suddenly fell down on the floor. The police came and were ready to arrest the veteran for public drunkenness. "He's not drunk," Boren told the officers. "He's suffering from a relapse of malaria."[15]

If it was easy for the enemy to distinguish a forward observer by his binoculars, it was even easier to spot the radio operators. Wearing the heavy, cumbersome piece of equipment was like presenting a target. During the fighting from January 27 through January 30, another radio operator of the 135th Field Artillery Battalion was not as fortunate as Corporal Josh. Pvt. Robert S. Veira was mortally wounded immediately after transmitting his forward observer's instructions to the FDC.[16]

By the thirtieth the 129th Infantry had captured Fort Stotsenburg while other elements of the division held most of Clark Field. The next day the bloody struggle to take Top of the World began. During the day, four different artillery battalions massed their fires to neutralize and destroy well-entrenched enemy positions, allowing the attacking 129th Infantry to advance and secure the area on February 2.[17]

On January 27 Beightler sent the 148th Infantry advancing toward Manila.[18] On February 1, as it was approaching Plaridel, about twenty-five miles north of the city, it came within about four hundred yards of the enemy's first line of defense. Here a Japanese garrison of five companies fought savagely in an effort to hold the critical road junction there. As the 1st Battalion moved forward, the Japanese fired a heavy concentration of mortar fire to repel the attackers. The next morning the Americans struck back with an intense artillery barrage. The coordination with the infantry assault was near perfect. Riflemen from Company A, 148th Infantry began moving out just as the shelling was ending and were able to push forward. Shortly before dawn on February 3, the Japanese withdrew.[19]

The forward observer with the 148th Infantry made very precise adjustments for the leading platoon of infantrymen, hitting every enemy strongpoint until shell fragments began falling among his own troops.[20] The observer may have been Lt. John S. Wallace of Tacoma, Washington. On February 2, Wallace was with the leading elements of troops assaulting Plaridel. At his post three hundred yards from the Japanese, he remained in position despite intense enemy machine-gun and mortar fire. The extreme accuracy of his adjustments neutralized the Japanese emplacements, enabling the infantrymen to advance while meeting only slight resistance.[21]

Wallace's situation exemplifies many similar ones that occurred during the war in which a forward observer on the ground, working from the most advanced position possible, would be able to provide more closely coordinated artillery support than one in the air. While aerial observation was frequently more advantageous because of the observer's broader perspective and greater mobility, the man on the ground was much more aware of the immediate situation as it unfolded and could react accordingly, often anticipating the infantrymen's most urgent requirements.

After clearing the Japanese from Plaridel and securing the road junction on February 3, the 148th Infantry continued their advance toward Manila. Two battalions of the 145th Infantry were the first Americans to reach the Manila suburb of Polo-Malinto the next day.[22] Although the 37th Division had received some limited training in urban fighting, most of their combat experience had been in a largely uninhabited jungle environment. The densely populated city before them posed a severe challenge to General Beightler, with his preferred heavy reliance upon artillery in battle, regarding how to avoid civilian casualties while fighting an enemy who had deliberately deployed troops among residents. Using artillery to support infantry in Manila while avoiding civilian casualties would be extremely difficult, if even possible.

In a deliberate effort to avoid civilian casualties and destruction to property, the Sixth Army placed stringent restrictions upon artillery fire during the early phases of the battle for Manila. Artillery fire was confined to observed fire only upon specific targets such as Japanese gun emplacements. Airstrikes had similar limitations until General MacArthur suddenly decided to halt them altogether.[23] Consequently, 37th Division artillery played a limited role in the first days of the fighting.[24]

On February 7, the guns fired a heavy barrage on enemy emplacements as the 3rd Battalion, 148th Infantry began crossing the 150-yard-wide Pasig River in the vicinity of the presidential palace. This was met by uncharacteristically heavy Japanese mortar and artillery fire.[25] Meanwhile, the first two battalions of the 129th Infantry had crossed the river and begun advancing toward Provisor Island. At 8:00 A.M. on February 9, assault boats carried the lead company of the regiment across the Estero de Tonque. After land-ing and establishing a beachhead, the Americans were met by heavy Japanese artillery, mortar, and machine-gun fire. Again, with the aid of strong artillery support, the infantrymen broke enemy resistance and two days later took control of the island. By February 10 the regiment had cleared most of the Pandacan District but encoun-tered strong Japanese opposition at Paco Railroad Station and the buildings of Concordia College and Paco School. Fire of the 136th and 140th Field Artillery Battalions almost entirely destroyed the buildings but saved the 148th Infantry many casualties.[26]

Although the Americans had used their artillery in Manila prior to February 7, higher headquarters had imposed restrictions. Because the Japanese had turned almost every large building in the city into a defensive position, the attempt to clear them without artillery support became a costly undertaking. Between February 7 and 10, the 148th Infantry suffered more than 500 casualties and was understrength by about 600 men. The 145th Infantry had also experienced heavy casualties over the same period of time and was about 700 men understrength.[27] As the division's casualties began to climb, General Beightler wanted the restrictions on field artillery lifted to protect his infantrymen. The attempt to avoid the destruc-tion of buildings would have to be abandoned if Manila were to be captured without the destruction of the 37th and 1st Cavalry Divi-sions. Lt. Gen. Oscar Griswold, XIV Corps commander, gave his consent to use artillery against all Japanese strong points except for public buildings, such as churches and hospitals, that would likely contain civilians.[28]

During the fighting of February 12–22, the 37th Division fought a war of attrition aimed at reducing Japanese positions east of Intra-muros, including several municipal buildings. Daily casualties were never particularly high, but they kept increasing. Beightler conse-quently stepped up the use of artillery.[29]

After rebuilding the bridges across the Pasig, the general moved artillery pieces, tanks, and tank destroyers into southern Manila to clear the way for the infantrymen. On February 12 two members of MacArthur's staff, Brig. Gen. Bonner Fellers and Col. Andres Soriano, a wealthy Filipino, visited the 37th Division's forward headquarters to observe the fighting. Soriano reportedly owned a number of buildings in the capital. Shortly afterward, Fellers reported to the supreme commander that Beightler was "unneces-sarily destroying the city." MacArthur then posted an order limiting the maximum size of any artillery used to 37 millimeters, a shell large enough to stop a light tank but not large enough to effectively penetrate concrete emplacements. Yet Beightler was troubled by the recent combat deaths of two officers, whom he knew personally, and the division's climbing numbers of casualties overall. Upon hear-ing the order, he told General Griswold that he "would rather be relieved of command than carry out that order." Within a few days, MacArthur had reversed his directive.[30]

Beightler's artillerymen came to realize that even 105-mm how-itzer shells had little effect against government buildings designed to withstand earthquakes. During this second half of the Battle of Manila, the division used a combination of fires to combat the Japa-nese. The batteries using smaller 105-mm howitzers were used indi-rectly from a distance, while several batteries of 155-mm howitzers of the 136th Field Artillery Battalion were moved within sight of their targets.[31]

The last major Japanese stronghold to be taken in the capital was Intramuros, on the south side of the Pasig River. Built in the sixteenth century in the manner of a European walled city, Intra-muros had a circumference of two and a half miles and walls forty-feet thick at the base. Within this perimeter, the Spanish had built Fort Santiago. The walled city also contained several tunnels. The side facing the river was open, the only portion of Intramuros *sin Muros*.[32] To reach any of the other three sides, attackers would have to cover a wide expanse of open land.

American intelligence was aware that among its estimated 13,000 Japanese defenders was a large civilian population. Figuring that such a large number of enemy troops would be able to offer strong resistance to any attempt to cross the river, planners with the 37th Division's G-3 section decided to make the initial attack from a beachhead east of Intramuros and then, shortly after, follow up with an amphibious assault. The plan was for the 129th and 145th Infantry Regiments to keep the Japanese occupied on the east while the 148th Infantry made the crossing.[33]

General Beightler wanted to bombard the objective as heavily as possible prior to the attack to minimize his casualties. He also asked for an aerial bombardment prior to the infantry assault, but

General MacArthur adamantly denied his request, citing the likelihood of inflicting heavy casualties on the civilians within the walls.[34] In a series of reports published after the war, MacArthur indicated that he decided against using bombers because he knew many civilians were there. The Americans instead broadcast a plea to Japanese forces to surrender or release the civilians inside, but they received no answer. Thus, the planned attack began.[35]

As early as February 17, American artillery began to bombard the outer walls for the purpose of breaching them and eliminating defensive positions along them. The initial barrages included indirect fire from ranges as far away as four and half miles as well as direct fire from as close as 250 yards. By the morning of the scheduled infantry assault, breaches had been blasted in portions of the east and north walls, creating openings among the fractured rock to enable attacking troops to scramble through.[36]

During the night of February 22–23, the division brought twelve 105-mm and six 155-mm guns to within a short distance of Intramuros on the north and east. Here the division used twentieth-century technology with medieval tactics, employing direct fire to expand existing gaps in the walls and to make additional openings for the infantry. The rest of the division artillery and other units provided indirect fire throughout the night before and during the infantry assault. This saturation of all points of the attack eliminated obstacles such as minefield and barricades in the path of the infantrymen. At 7:30 A.M. on February 23, all thirty-six direct-fire field pieces brought in the night before opened fire on Intramuros for one hour, supplemented by the guns of a tank battalion and cannon companies.[37] Thus, while direct fire was used to a limited extent, it was used in conjunction with indirect fire, which forward observers inside of buildings adjusted on a different set of targets.[38]

The initial plan of attack called for the 129th Infantry to make an amphibious crossing of the Pasig River and attack Intramuros from the north while the 145th Infantry was to attack simultaneously from the northeast. During this time, the 148th Infantry would provide fire support from the north bank of the river. The one-hour barrage that preceded the infantry assault was probably one of the most well coordinated and destructive of the entire Luzon campaign. Each artillery unit received a designated target area that overlapped the others so that the barrage covered virtually the entire fortress.[39]

Nine battalions of field artillery, including 240-mm howitzers, one battalion of tanks, and one battalion of tank destroyers, would take part in a one-hour "time on target" mission.[40]

The 37th Division's artillery poured roughly 185 tons of high explosives on the walls of the fortified city.[41] The resulting cacophony was so loud and intense, with the gunners only sixty to one hundred yards away from their target areas, that telephones and radios at the guns were virtually useless. Instead, the crews employed a system of visual signals to relay commands.[42]

As late as 8:15 A.M., the heavily barricaded Quezon Gate continued to withstand all efforts to destroy it. At this point the tank gunners of the 145th Infantry switched from other targets and concentrated on the gate, blasting a gaping hole in it and facilitating the attack of the 145th's infantrymen.[43] As the last shells were still in flight, the 2nd Battalion, 145th Infantry attacked the east wall at the same time that the 3rd Battalion, 129th Infantry was approaching the north wall in assault boats. As a result neither group met any significant enemy resistance until they were inside Intramuros.[44]

Overall, the entire operation went very well. By 10:00 A.M. on February 23, Beightler declared the assault a success. Although considerable fighting remained to be done, by then he was confident that, having reached the interior of the walled city, it was unlikely his riflemen would be driven out.[45]

Casualties for the 37th Division were twenty-five killed and 265 wounded, while an estimated 2,500 Japanese died, half from artillery fire. At one point, as the attack was losing momentum, the Japanese deliberately released nearly 3,000 civilians from San Augustin and Del Monico Churches. Their sudden appearance forced the Americans to hold their fire until these people were out of harm's way.[46]

Beightler believed that the comparatively easy victory had been achieved by the artillery bombardment. Those Japanese not killed were so dazed that they were unable to continue fighting with any degree of coordination.[47] By the evening of February 24, the 1st Battalion, 129th Infantry had eliminated all organized resistance in Intramuros.

After that the remaining Japanese held out in various government buildings, all built from reinforced concrete and designed to withstand earthquakes. A heavy artillery bombardment of these began on March 24, with the infantry scheduled to assault them two days later. Again, as with Intramuros, the 37th Division employed their combined-arms tactics well, using direct and indirect artillery fire in conjunction with well-timed infantry assaults while unavoidably destroying much of Manila's infrastructure.[48]

The 155-mm howitzers of the 136th Field Artillery Battalion, firing directly at ranges from 150 to 800 yards, yielded the most effective results, breaching holes through the concrete during the preattack bombardment.[49] All gun sections of the battalion took part, operating their weapons while under heavy fire from enemy small arms and mortars. This direct-fire mode had one unanticipated consequence. The nearby 1st Cavalry Division reported that shells were reaching its area. An immediate investigation determined that 155-mm rounds fired directly at the Legislative Building were passing through the structure and landing in the 1st Cavalry's sector. General Beightler quickly called for a cease fire until adjustments could be made.

The basic procedure used during the attack on the municipal buildings was to pound the Japanese held up inside with a heavy artillery and then send infantrymen forward with flamethrowers and incendiary grenades.[50] During this phase of the battle, the 136th Field Artillery Battalion lost more battery personnel than forward observers, with one officer and four enlisted men killed and six officers and fifty-three enlisted men wounded. This was because they were so unusually close to the enemy.[51]

Maj. John Gordon wrote that the precision required by direct firing during the fighting in the streets of Manila had forced the artillery to move very close to its targets. Because these enormous, well-constructed buildings were impervious to tank or even 105-mm artillery fire, the heavier field guns had to be used. Despite the need to employ direct fire, Gordon noted that "the vast majority of the ammunition expended was still for the indirect fire, accounting for more than 90 percent of all the ammunition fired."[52]

On February 26 the division launched its infantry assault, and one by one the last three strongholds fell. Although the honor of being the first American army unit to enter Manila belonged to the 1st Cavalry Division, the 37th Division had the dubious distinction of ending the last organized enemy resistance in the city, this within the Finance Building. On March 3 General Griswold reported to General Krueger that the Manila area was secured.[53]

The 37th Division's experience in Manila was, in part, a throw-back to the way in which artillery had fought earlier battles. Field guns in plain view of the enemy meant that gun crews were sure to suffer casualties. The way in which artillery was used to reduce strongly constructed concrete buildings and the thick walls of Intra-muros was even reminiscent of medieval warfare, when armies laid siege to a walled enclosure and fired heavy cannons to breach them. Direct fire did not require forward observers to hit the tar-get, although someone observing the results might be able to direct subsequent shots more effectively. Despite this, forward observers remained busy in Manila directing neutralizing fire to destroy tar-gets of opportunity and to keep the enemy disorganized.

The forward observers of the 1st Cavalry Division were also actively involved in the fight for their sector of the city. Lt. Charles E. Doesburg of Glencoe, Illinois, was with 82nd Field Artillery Battal-ion. At Santo Tomas he climbed a tower just as enemy artillery and mortar shells began to fall so he could adjust counterbattery fire on Japanese guns nearby. While there, enemy shells hit the tower nine times, killing more than thirty and wounding nearly one hundred people nearby. As others around him sought cover, Doesburg stayed in his exposed position, directing accurate artillery fire.[54]

With American troops now in control of Manila, most of west-ern Luzon, and the central plains, the Japanese began to withdraw to the more mountainous areas of the north and east. With the advantage of holding the high ground, they would try to hold until such time when they could eventually drive the Americans into the sea. On March 25 the 37th Division received new orders assigning it to Baguio. Three days later the 129th Infantry was attached to the 33rd Division with the mission of heading east along the Naguilian Road, preceding the rest of the 37th Division.[55]

By then the remainder of the division had already departed from Manila and begun the advance toward Baguio. On April 18 the 6th Field Artillery Battalion experienced a serious problem in com-mand as it was providing support for the 129th Infantry. Lt. Col. Stewart L. Brown had driven forward to observe the effect of his unit's counterbattery fire. A Japanese 150-mm shell scored a direct hit on the observation post where Brown was watching, killing him and two others by his side. Lt. Col. Chester A. Wolfe then assumed command of the 6th Field Artillery.[56]

On April 14 the 148th Infantry passed through the 129th Infantry east of Monglo. Two days later it was approaching the Irasan River gorge about five miles west of Baguio. After destroying the bridge on Route Nine, the Japanese took up defensive positions in a series of ridges overlooking the gorge. The battle for Irasan began on April 17 and lasted four days.[57] En route, the 148th Infantry met scattered Japanese forces, which fought back with artillery and mortar fire. With support provided by the 140th Field Artillery Battalion, forward and aerial observers adjusted counterbattery fire to quickly end the enemy threat.[58] Finally on April 21, the 1st Battalion, 148th Infantry secured the rest of area near the Route Nine bridge. The 129th Infantry passed through the 148th on April 22 as the division continued its advance.[59]

On the evening of April 23, as elements of the 129th Infantry approached Baguio, they met strong resistance at the village cemetery.[60] Overnight, the Japanese launched a counterattack led by tanks against the 129th's position. As the 6th and 136th Field Artillery Battalions laid down a heavy protective barrage, the infantry and supporting tanks repelled the attack.[61] The next morning, after an artillery barrage, the 129th overran the cemetery, and by the evening of April 26, the division had taken Baguio.[62]

As he had done in previous campaigns, Beightler made extensive use of artillery during the Baguio operation. But readily available air support and a shortage of artillery shells forced him to limit this usage to well-defined targets rather than harassing and interdictory fire. By now tactical air support had greatly improved, and he began to step up his use of it. By the end of the Baguio campaign, aircraft could safely bomb enemy targets within 400 yards of his front lines, sparing the infantry many casualties while greatly facilitating its progress.[63]

Effective counterbattery fire, the destruction of Japanese artillery pieces by American howitzer batteries, played an important role in the successful execution of combined-arms tactics on Luzon. The 37th Division came under more counterbattery fire than it had in the previous two campaigns, yet the obvious inequality between the Japanese artillery arm and its American counterpart persisted. This was partially due to the efforts of forward-observation teams. During the period from April 26 to May 4, Lt. William F. Sullivan of Lakewood, Ohio, while under heavy enemy fire, directed artillery strikes resulting in the destruction of three Japanese field pieces, sparing the infantrymen many casualties and enabling them to advance.[64]

As the Baguio operation was winding down, Sixth Army planners decided to send the 37th Division up Route Five through the Cagayan Valley, formed by the Cagayan River flowing 260 miles amid a series of high mountain ranges, the last Japanese stronghold on Luzon.[65] This was the last sustained-combat operation the 37th Division would fight and was characterized by its mobility and concentration of firepower. The 129th Infantry Regiment spearheaded the advance on May 31, and for the next four weeks, the division's three infantry regiments repeatedly leapfrogged each other. Stanley Frankel has described the massive firepower available to the 129th Infantry: "Overwhelming power was the keynote of the entire operation."[66]

Later General Beightler would refer to this chapter in the division's history as its blitzkrieg. It was certainly a good example of *Bewegungskrieg* (mobile war). By now Japanese forces had become fragmented and were unable to recover. In less than four weeks after leaving Santa Fe, the 37th Division had advanced 225 miles to the north, ending the last phase of liberating Luzon. In its wake were large quantities of destroyed enemy material and many dazed Japanese soldiers wandering among the hills.[67] Noting the rapidity of the advance, a forward observer from the 6th Field Artillery noted that the guns barely had time to fire a few rounds for registration before the infantry was on the move again. He also claimed that it was even difficult for aerial observers to keep pace.[68] This posed a challenge to both in maintaining control of artillery fires.

By June 10 the Japanese had pulled back to the vicinity of a wrecked bridge at the entrance to Orioung Pass, and as lead elements of the 145th Infantry advanced, the Americans quickly incurred heavy casualties before halting abruptly. After the 135th Field Artillery fired a heavy concentration, soldiers of the 2nd and 3rd Battalions moved forward only to stop again after encountering heavy automatic-weapons fire. Another envelopment to the north also came to a halt, and during midafternoon, all attacking units fell back while the 135th and 136th Field Artillery, joined by the 251st Field Artillery, fired a massive barrage for nearly two hours. After the firing stopped, the 2nd Battalion, 145th Infantry advanced and by nightfall had gained control of the high ground at the entrance to the pass.[69] Despite using heavy artillery and mortar concentrations in conjunction with airstrikes, it took three days to clear the pass. Although the 37th Division had moved quickly, nearly two weeks of the campaign had passed before the Buckeye Division actually entered the Cagayan Valley, reaching Cordon on June 13.[70]

To compensate for its weakness in field artillery, the Imperial Japanese Army relied heavily on mortars. These light weapons are comparatively much more mobile than field guns but can deliver a devastating amount of firepower and destruction. Although a mortar also emits a muzzle flash when fired, it was quite difficult for forward observers to detect them because the extent of the entire weapon is a tube and a base plate, making it relatively easy to conceal. Major Rogers described this as the enemy's most consistently effective weapon on Luzon, adding that countering fire was problematic: "Destruction of these mortars has been principally due to alert forward observers and aggressive infantry action. Location by sound, flash, or air observation is extremely difficult."[71] Although aerial observers gained a much wider, expansive view of the battlefield, the Japanese were so good at the art of concealment that ground observers on Luzon were often able to spot targets unseen from the sky. Depending upon the terrain, it could work either way.

An incident illustrative of this occurred at Orioung Pass when a tank column was supporting units of the 145th Infantry. The column consisted of light and medium tanks as well as radio cars from the Signal Corps. Riding along were an Army Air Corps liaison officer and a field-artillery forward observer, who maintained radio contact with an observer in the air. Thus when he noted something the aerial observer did not, he could relay the target information to the flyer, who would then tip off the tank crew nearest to the enemy location. This artillery-airplane-tank team worked very well.[72] It also demonstrated that forward observers could acquire targets and control fires, even if their instructions were relayed through multiple channels and even if armor, not artillery, ultimately provided the supporting fire.

On June 17 Lt. William Gaylord of Lacomb, Oregon, a forward observer with the 140th Field Artillery Battalion, was killed in action. On that day the 2nd Battalion, 148th Infantry encountered the Japanese at the Cagayan River just south of Naguilan. The river was not fordable, and the enemy put up a stiff defense with artillery and machine guns.[73] Gaylord heedlessly exposed himself to enemy fire to reach an optimal vantage point from which to direct artillery. After transmitting his initial instructions, he was struck and killed by machine-gun fire. Yet from the accuracy of his directions, a heavy barrage routed the Japanese, enabling the infantry to reach its objective with far fewer casualties.[74]

Shortly afterward, the 148th Infantry passed through the 145th, and on June 23 the 129th passed the 148th. By June 26 the men of Company F, 2nd Battalion, 129th Infantry had made contact with the converging paratroopers of the 511th Parachute Regiment, 11th Airborne Division.[75] The next day the 129th Infantry entered the city of Aparri, ending both the Cagayan operation and the Luzon campaign.[76]

Frankel attributes much of the 37th Division's success in the Cagayan Valley to its ability to maintain its momentum, keeping the Japanese off balance and forcing them to fight small but strong delaying actions. Well-hidden enemy positions meant small-unit actions rather than the standard two-battalion frontal assault were the norm. As a result it was extremely important to keep the supply lines open at all times.[77]

The rapid displacement to Cagayan Valley had created a strong challenge for the artillery to keep up and within range of the leading elements of the infantry. The advance could only be made on narrow and winding roads, which left the division severely extended.[78] Service batteries had to haul ammunition as far as 230 miles over dirt and gravel roads that crossed mountains and bypassed destroyed bridges. In the valley artillery had flat, open areas to deploy along the roads, while the enemy held the high ground on its flanks.[79]

Major Rogers noted that by maintaining its rapid pace, the artillery had been in position to make the advantageous use of its fire as the locations of various Japanese emplacements became known. In a count of enemy equipment known to have been destroyed by artillery fire, Rogers listed "524 vehicles of all types. Most of these were destroyed when artillery observers caught hundreds of enemy vehicles jammed on Highway Four between Bagabag and Kiangan in the Jap's mad flight."[80]

The successful drive through Cagayan Valley virtually assured American victory on Luzon, though not total Japanese defeat. There remained in excess of 50,000 Japanese troops on the island by the end of June. More American soldiers were to die during July and the first half of August, but never at the same rate as earlier. The final assignment of the war for the 37th Division was to keep in check the 13,000 enemy troops scattered in the mountains east of the Cagayan Valley. Although the fighting in this area consisted of isolated skirmishes, the division lost another fifty killed and 125 wounded in the last six weeks of the war. On August 10, after listening to a radio broadcast telling of Japan's intentions to surrender, Beightler immediately ordered his troops to cease aggressive action in order to avoid further casualties within his division.[81]

Luzon represented the 37th Division's first experience fighting in a populous region, with much land under cultivation and a large population center in a major urban area. These conditions presented new challenges for the effective operation of infantry and artillery using combined-arms tactics. They generally met these challenges very well.

For the first time, the 37th Division could maximize its capacity for aerial observation, sometimes in conjunction with ground observation, since much of Luzon's topography included open plains and cultivated land. The division's experience was that aerial observation was superior for targets in open terrain or in motion. The destruction wrought by artillery on the Japanese in the Cagayan Valley underscores this point. Ground observation, however, was essential for revealing concealed positions, which the Japanese used masterfully. The observer on the ground could also use sound to spot the enemy, while the aerial observer could not.

The latter stages of the campaign also demonstrated the improvement in coordination between infantry and its air support that had developed by point in the war. Frankel and others have commented how tactical air support had not yet reached its full potential in the earlier island campaigns. In a postwar report General Beightler admitted that during the Cagayan operation, his infantry had benefited not only from the excellent coordinated effort of its air support but also from all available combat arms, destroying all enemy artillery it encountered and neutralizing about seventy Japanese tanks.[82]

The major lesson of Orioung Pass was that the forward observer could direct other fires in addition to artillery, simultaneously expanding his destructive capabilities while enlarging his respon-sibilities. He could find, detect, or acquire targets and could then conduct and adjust the fires of other service branches and combat arms. Fire support did not have to come from the field artillery alone; air or armor, the Marine Corps or navy could provide it as well. This was today's joint-fires concept in its infancy. At Orioung Pass ground observers acquired targets and aerial observers con-ducted and adjusted fires, sometimes using artillery, sometimes using armor.

The 37th Division's most challenging use of artillery on Luzon was in Manila during the assault on Intramuros and the subsequent attacks on the concrete government buildings. The effort to launch an infantry assault on a sixteenth-century city with walls twenty-feet high and from twenty- to forty-feet thick relegated modern field artillery's indirect fire almost to the level of medieval siege tactics, demonstrating that some situations in modern combat still require direct artillery fire. It also negated the advantages of modern com-munications technology. Gun crews and the FDC had to communi-cate with hand signals because they were unable to hear each other over the roar of the guns and explosions.

An estimated 100,000 civilians died during the battle for Manila.[83] While the purpose here is not to explain how that hap-pened, it is interesting to note that some hold Beightler's extensive use of artillery responsible for the extent of civilian casualties and destruction to the city.[84] While no one knows how many civilians American gunfire killed unintentionally, what seems apparent is that the Japanese deliberately massacred thousands of Filipinos and made no effort to provide for the safety of others during the course of combat, at times virtually using them as hostages.[85]

The story of collateral damage in Manila reminds us that, espe-cially in this nuclear age, what is required to defeat an enemy in war may result in a high number of civilian casualties and the destruction of property, regardless of the sophistication of weaponry and avail-able firepower, and that the great levels of mass destruction achieved may not necessarily be confined to the enemy. The objective of war, however, is to destroy that enemy's ability to resist, and the 37th Division certainly realized that goal in the battle for Manila.

| Initiation to Combat

The Saar Basin

The initial experiences of the U.S. Army's 87th "Golden Acorn" Division in Europe demonstrated how much more parity existed between the Americans and the Germans at the tactical level than existed between the combatants in the Pacific War. Hitler had spread the Wehrmacht too thin by forcing it to fight a two-front war, and by the end of 1944, Germany was also experiencing a decline in crucial resources. By then the Germans had well-rounded experience using combined-arms tactics in mobile warfare that the Japanese Imperial Army never had. The German army was also extremely good on the defensive, where it utilized combined-arms warfare to its fullest advantage as the 87th Division began its offensive operations.

In December 1944 the 87th Division consisted of the 345th, 346th, and 347th Infantry Regiments; Headquarters and Headquarters Battery, Division Artillery; 334th Field Artillery Battalion (105-mm); 335th Field Artillery Battalion (155-mm); 336th Field Artillery Battalion (105-mm); 912th Field Artillery Battalion (105-mm); 87th Reconnaissance Troop (Mechanized); 312th Engineer Combat Battalion; 312th Medical Battalion; 87th Counterintelligence Corps Detachment; Headquarters Special Troops, Headquarters Company, 87th Infantry Division; Military Police Platoon; 787th Ordnance Light Maintenance Company; 87th Quartermaster Company; and the 87th Signal Company. Armored support came from the 735th Tank Battalion (attached February 1–March 9 and

March 15–May 9, 1945), 761st Tank Battalion (attached December 20–23, 1944; January 1–15, 26–February 1, 1945), 607th Tank Destroyer Battalion (attached February 3–March 6, 15–May 9, 1945), 610th Tank Destroyer Battalion (attached December 14–22, 1944), 691st Tank Destroyer Battalion (attached December 22–24, 1944; January 8–26, 1945), 704th Tank Destroyer Battalion (attached December 17–19, 1944), 811th Tank Destroyer Battalion (attached January 26–28, 1945), and the 549th AAA Automatic Weapons Battalion (attached December 24, 1944–May 9, 1945).[1]

There were great similarities between the Germans and Americans and their general artillery usage. Standard field pieces were fairly evenly matched, and both sides employed forward observers at or near the front to great advantage. There were also similarities in communications, with the use of radios and telephone lines.

After a short stay at Metz, the 87th Division went into the line December 11, 1944, with orders to cross the German border. Unknown to them at the time, this meant breaching portions of the Siegfried Line, a series of interconnected defensive fortifications running along Germany's western border from Switzerland to the Netherlands. Because Hitler's army was fighting to keep an enemy from invading its homeland, this would be a difficult struggle, pitting the Golden Acorn Division against an experienced, sophisticated foe capable of fighting a complex and ferocious form of war.

One point often overlooked is that at the onset of America's entry in World War II, many National Guard units were better trained and prepared for combat than were the members of the regular army.[2] Young men of the Depression era frequently entered guard service after leaving school since it entailed less of a commitment than the army and represented a way to earn an income in a time when jobs were scarce. After Pearl Harbor, a number of regular-army divisions had to build their infantry regiments and field-artillery battalions around a tiny cadre of career soldiers. As a result, many draftees and new enlistees were thrown together as individual divisions were activated. Because many of the guardsmen had joined right out of high school during the last years of the 1930s, a large number of them had trained and worked together for years, not months, even before the guard was federalized. Battery C, 135th Field Artillery was one example.

On December 15, 1943, the 87th Infantry Division reentered active duty. By that time the 37th "Buckeye" Division had fought on

New Georgia and had begun the Bougainville campaign. The 87th was a typical triangular division, typical too of those American units that joined the war in Europe after D-Day. During its five months' service on the continent, it experienced steady combat, some of it quite intense. The division used its artillery well, and a study of its forward observers provides a good basis for comparing their collective experience in Europe with that of the forward observers in other army units in the Pacific.

Arriving at Fort Jackson, South Carolina, in early 1944, the division trained there prior to its deployment that fall. By early December the 87th Division had reassembled in the vicinity of Metz, France.[3] At that point the U.S. Army had two years of combat experience in the war, and American forward observers had gained even more practice applying their new techniques in battle. Despite this, most units new to the European theater were generally assigned to less intense areas of fighting until they had gained some exposure to combat. This explains why the Golden Acorn Division began the war at the old fortress complex in Metz, still occupied by the Germans at that time. Although the commander of the city had surrendered on November 21, by the end of the month, the garrisons of four of the forts—Driant, Plappeville, St. Quentin, and Jeanne d'Arc—in the complex continued to hold out.[4] The 5th and 95th Infantry Divisions had suffered heavy casualties trying to subdue these strongly fortified positions.[5] In early December the 87th arrived in the outlying region. On December 6 at 10:55 A.M., Battery A, 334th Field Artillery Battalion opened fire on Fort Jeanne d'Arc. PFC Donald McCabe of Philadelphia, Pennsylvania, pulled the lanyard.[6] This was the first shot fired by the 87th Division in World War II.

Because it was the unit's *first* shot, the division commander, Maj. Gen. Frank L. Culin, wanted the shell casing as a souvenir. But the enlisted men of the gun crew were determined that one of their number should have that memento. Six decades later Sam Delli of Lyndhurst, Ohio, recalled how they outwitted the general. The men saw Culin approaching, but before he reached them, they had fired a second shot. Delli was the number-three man on the howitzer. Delli kicked the casing from the first shot under the gun, and the general unknowingly walked away with the one from the second round. After Culin left, McCabe got the prized souvenir.[7]

The Golden Acorn's stay in Metz was only for about a week, and its artillery experience gained there was not much more than practice. Its real initiation to combat now lay close at hand. "On 10 December . . . orders were received" for the 346th Infantry "to relieve the 104th Infantry Regiment of the 26th Infantry Division at dawn 11 December 1944. The next day found the attack shifting in the direction of Rimling, France."[8] Meanwhile, the 347th Infantry had moved into the line and began its initial assault the same day, advancing toward the French village of Obergailbach. After meeting bitter opposition, the 347th took the town and the heights overlooking the Blies River on December 15.[9] This brought the division to within a few hundred yards of the German border to the north.

On December 13 the 912th Field Artillery also carried out its first combat mission, firing on German troops in the vicinity of Obergailbach. The battalion's first three casualties were forward observers. "Late in the afternoon of 15 December 1944 . . . Capt. Leonard Harding, who was returning to his Battery after an assignment as a forward observer, was hit in the shoulder by an 88-fragment. The very next day . . . Capt. William F. Botkin and his radio operator, PFC Raymond Johnson, both of Battery B were killed by an enemy artillery shell which landed in their foxholes."[10] Also on the fifteenth, "the 345th's 1st and 3rd Battalions took over from the 346th Infantry's 1st and 2nd Battalions in the vicinity of Rimling, France," a few miles southeast of Obergailbach.[11] From here the 345th prepared to cross the border and invade Germany near the beautiful village of Medelsheim, where that regiment would engage in its first large-scale combat.

The battle that took place there on December 16 illustrated some of the problems yet to be worked out in the successful execution of combined-arms tactics, particularly with regard to command and communications functions. The Golden Acorn Division was part of the Third Army, and as it went into the line in northern France, it represented the extreme right flank of Lt. Gen. George S. Patton's command. To the right of the 87th were elements of the Seventh Army. After the Germans opened fire with heavy artillery, tank, and machine-gun fire on the 3rd Battalion, 345th Infantry, forward observers immediately found targets but were unable to obtain permission to conduct artillery fire until the request went

As Company L emerged from this woods on December 16, 1944, it came under attack. The border of France is just beyond the woods, while Medelsheim is to the left-rear. Photo by author.

up the chain of command and the boundaries of the two American armies could be more precisely ascertained. This took several hours. In the meantime the infantrymen of Company L paid the price.

Starting from a position near Rimling, France, the 3rd Battalion, preceded by two artillery barrages, jumped off at 5:00 A.M. It headed toward the Mertzenwald-Baumbusch Forests, which straddle the German-French border. The men advanced almost a mile, and by daybreak Company L occupied both Merzenwald in Lorraine and Baumbusch in Germany.[12]

At this point Col. Douglas Sugg, commander of the 345th Infantry Regiment, ordered Capt. Howard Wall, commanding officer of Company L, to continue the advance. Wall objected strongly, telling his superior that the right flank remained totally unprotected and that the company should not move forward without contact from the 87th Reconnaissance Troop, purportedly on the company's flank. Sugg replied that the men were there and to move out. "To which Capt. Wall boldly but accurately replied: 'The hell they are,' [and] hung up his field phone." Then the captain ordered his men to move out.[13]

Wall undoubtedly wanted confirmation of the 87th Recon's location at that time so he could request an artillery barrage to precede the 3rd Battalion's advance without endangering friendly troops. But Sugg's false affirmation that the unit remained in the targeted area ruled this out. Wall, who was later killed near Olzheim, Germany, reluctantly carried out his orders.[14]

Just as PFCs Louis F. Stein and Raymond Kline, the two leading scouts of Company L, were nearing the top of a ridge overlooking Medelsheim, a German tank hidden in the woods opened fire, killing them both. The burst signaled German artillery to commence firing. The barrage caught the men of Company L in the open on a hillside.[15] Another possible foul up that day involved the division's armor support. Sam Jones of Company L, who witnessed Stein's death that morning, observed: "As we sit around waiting for daylight, I was shocked to see four medium sized tanks sitting below the crest of the hill behind us with fire shooting out of their exhaust. We never had tank support before and didn't know what to think about it. We moved out at daybreak and never saw or heard about the tanks again."[16]

Along to provide forward observation for the 3rd Battalion were three members of Battery A, 334th Field Artillery Battalion: Lt. James R. McGhee of Mount Vernon, Illinois; PFC Gayle Bricker, Jr., from Sarver, Pennsylvania; and Tech 4 Donald L. Walker of Alliance, Ohio. McGhee remembered that the Germans started shelling them from their right-rear while enemy machine guns in Medelsheim to the right-front placed fire across the ridge.[17] Walker was carrying the radio for the party, and at this critical moment, it became inoperative.[18] McGhee then told Bricker and Walker to take shelter in a large pile of rocks in a gully.[19] The infantry radio was working without problems, so the lieutenant accompanied Captain Wall and Sgt. Paul Cutler, the captain's radio operator, in a frenzied rush to a building atop the ridge so he could maintain radio contact with battalion artillery.[20]

Don Walker, Battery C, at Camp Shelby, Miss., 1941. Photo courtesy of Donald L. Walker.

After the first German tank had opened fire, "all hell broke loose." As Wall, Cutler, and McGhee reached the top of the hill, they approached a small brick shed from the rear. Using the bazooka to open the door, they punched a hole in the wall and crawled inside, where they had a good view of Medelsheim. From the shed they could see inside the steeple in the village's lone church. With binoculars they saw the motion of observers or snipers within the steeple. "There was enough traffic entering and leaving Medelsheim, especially along the east–west roads north of the church . . . to suggest that here was a vital 'motor pool' or certainly part of the larger network for the Germans attacking in the Ardennes on that very day."[21] (This was a logical conclusion: today the driving time from Medelsheim to the Schnee Eifel, where the Ardennes Offensive began that morning, is a little over two hours.) McGhee obtained quick approval to bring artillery fire on Medelsheim. After he successfully directed a time-on-target, the little village disappeared in a cloud of smoke, eliminating the machine guns but not the artillery shells, which continued falling on the Americans.[22]

One of the wounded, Donald F. Hoehle of Columbia, Missouri, remembered: "The Germans let us [Company L, 345th Infantry] advance considerably out into the open before they started laying 88mm fire on us. Then, when they had us pinned down, they began strafing the area with machine gun fire."[23] Hoehle continued: "It was apparent that we couldn't advance without relief from the artillery shelling, so the call went back for counter-battery fire. . . . The initial rounds fell short, with the result that we had incoming rounds from both enemy and friendly fire. Eventually the range was corrected, but the enemy fire kept coming in."[24]

Shortly after the machine-gun fire stopped, Lt. Charles Ehret was given the task of ordering the men of Company L, including the two members of Lieutenant McGhee's forward-observer party, to fall back while German shells continued to rain down among them. Bricker and Walker refused to leave, waiting for their own lieutenant to return. As the medics moved forward to try to move the wounded to a place of safety, others joined in to help, including Walker. For some of the wounded, there was not much that could be done. Walker noted the case of one infantry captain who had been stitched across the thighs by machine-gun fire. Later Walker wrote; "We propped him up against a tree, where he bled to death."[25]

Medelsheim's church and steeple as rebuilt after the war. Photo by author.

Division headquarters afterward noted this incident: "He [Walker] then aided in evacuating infantry casualties from under direct 88mm fire, remaining in the area until every casualty was removed."[26]

The continuing enemy artillery originated from Utweiler, to the southeast of Medelsheim. The entry made at 12:50 P.M. in the 87th Division artillery's S-2 "Record of Enemy Activity" on December 16, made by "Highwater," reported mortars and 88-mm howitzers in Utweiler, plus ninety infantrymen and one tank.[27] "Highwater" was the radio codename for one of the artillery officers of the 912th Field Artillery Battalion, under the command of Lt. Col. Joseph A Monn.[28] After the time-on-target, Wall asked McGhee to try to break up the tank activity to their front and ordered Lieutenant Ehret to move the rest of Company L, still under the barrage, off the hill.

Battery A's reconnaissance officer, Lt. Guy Allee of Burlington, Iowa, another observer, could actually see the enemy's muzzle flashes but was unable to obtain timely permission to commence counterbattery fire. Meanwhile, Company L continued to suffer numerous casualties despite the fact that McGhee had finally forced the tanks to his front to withdraw. He later wrote that it took Allee "almost four hours to get clearance to fire into the Seventh Army's zone and silence the enemy guns!" That incredible delay unnecessarily reduced Company L, in a single day, from a full-strength rifle company down to about 80 percent of full strength.[29] Whatever the actual number of its losses, in its initiation to combat, the unit was "blooded" in the truest sense.

Years later Guy Allee explained that a large part of this delay was because of the 87th's location on the extreme right flank of the Third Army; to the division's right was the Seventh Army. When the Germans opened fire with their self-propelled artillery, Allee could see the muzzle flashes and reported their locations. He received word that this spot was within the Seventh Army's sector and that he was prohibited from directing fire on that location. Calling again, he assured the FDC that he could plainly see the target. This request traveled from the battalion FDC to the 87th Division artillery's FDC. Lacking the authority to approve his request, division artillery relayed his request to 87th Division headquarters, which in turn forwarded it to 3rd Army headquarters, which then sent it to Seventh Army headquarters for approval. Not knowing the exact positions of all units, Seventh Army relayed it to the headquarters of the division believed to be in the target location. The request then went down through that division's channels until it reached one of its artillery battalions' FDC. After their people determined that none of their division's troops were in the designated target area, they re-routed approval back to the 334th Battalion's FDC through the same great circular route that it had previously taken. "After this was received, they did respond and with artillery fire I was able to put an end to their [the Germans] firing on our troops."[30]

The Seventh Army units in the line on December 16 included the 45th, 103rd, and 79th Infantry Divisions and the 14th Armored Division. But the towns in Lorraine they occupied lay considerably east and much closer to Karlsruhe.[31] The 45th Division's 44th Cavalry Reconnaissance Troop may have been the unit adjacent to the 345th Infantry's position that day. That division's history indicates that "the latter part of December found the troop billeted in

Mayerhof [*sic*], France maintaining contact between our division and the 87th Division on the left flank, and the 100th Division on the right flank."[32] By December 23 the 44th Cavalry contacted part of the 87th Division at Obergailbach.[33]

Other friendly troops in that area included the "Hellcats" of the 12th Armored Division. That unit's history places the 87th Division on the left and the 44th on the right to the 12th's front, where they converged to take over the line from the "Hellcats." At this point the corps boundary changed, and the 80th Infantry Division then relieved the 12th Armored.[34] Medelsheim is located just north of the French village of Erching. Thus, if the 12th Armored Division was positioned to the rear of the 87th on December 16, and if the 44th Division had reached Meyerhof that day, this would have placed Seventh Army units within six miles of the 345th Infantry during the battle. Yet on December 16 a circulated 87th Division S-2 "Record of Enemy Activity" report indicated that the 12th Armored may also have been somewhere in the vicinity of Medelsheim, where the Germans had set a trap.[35]

The trap was reported to be in the valley of the Bickenalb, no more than a small creek with a rather north–south course where it passes between Medelsheim and the hamlet of Peppenkum. As it continues north to the east of Seyweiler, the stream takes a more northeasterly course. The path the 3rd Battalion, 345th Infantry followed toward Medelsheim was roughly 1,500 yards east of the Bickenalb and somewhat parallel to it. The 87th Division Artillery received a copy of the 87th's report, but by then the trap had already been sprung. Company L's initiation to combat took a toll, but perhaps not as high as believed at the time.

During World War II, the U.S. Army used a form called a "morning report" to indicate the number of soldiers present in a particular unit each day and to record changes in the individual status of its men. Company L's morning reports for the early part of December 1944 show approximately 180 members, with no significant changes in total strength until December 20. But the record for that date lists the names of twenty-nine men who left the unit on December 16 and another seven who left the next day. Perhaps the morning reports are in error, but if they are not, they indicate that most of Company L survived the battle unscathed. Possibly, the last seven were not discovered until the following morning. Yet had critical artillery support arrived sooner, fewer casualties would have resulted. These reports give no indication whose wounds were fatal and whose were not; "SFW" (shell-fragment wound) appears as the most frequent type of injury.[36]

The 345th Infantry's regimental history lists by name and company, though not by date, all members of the unit who were killed in combat. It includes two of the same names appearing on Company L's morning report for December 20.[37] Yet another source, *Casualties: Saar Valley—16 December 1944*, lists by name and unit all members of the 345th Infantry killed or wounded on December 16. This record indicates that twenty-one men of the of the 345th Infantry died in action that day, including four men of Company L not listed on the company's morning reports. The regimental listing corroborates the deaths of these four, though without reference to any date. *Casualties: Saar Valley* also lists sixty-four men of the regiment wounded in action on December 16, including twenty-three members of Company L whose names appear on the morning report. The only names absent from the wounded are six listed on the morning report as "NBC" (non-battle casualties).[38]

Although the accuracy of the records themselves could be disputed, taken together, they seem to corroborate that Company L incurred the majority of the battalion's wounded on December 16 but not those killed. The assertion that the company was reduced to thirty effectives may or may not be entirely accurate. What is significant, however, is that the Third Army command structure somehow impeded the forward observer's ability to bring fire to bear on observed targets expeditiously. Someone along the chain of command did not acknowledge or pass along in a timely manner the firing request. As a result the Germans killed and wounded many Americans who might have remained unharmed with available close-fire support. McGhee and Allee deserve praise for maintaining visual contact with their targets for so long. McGhee's willingness to remain with the battalion despite the loss of his radio spared the company additional casualties.

On December 17 the 2nd Battalion, 345th Infantry passed through the 3rd Battalion and attacked in the direction of Seyweiler. Company F experienced a disastrous incident apparently involving friendly fire that resulted in significant casualties for both it and Company E. The battalion jumped off around 9:30 A.M. heading toward Medelsheim. Anticipating tank support that never showed, the infantry approached the German positions about 11:30 A.M. Two enemy tanks did appear, however, immediately putting machine-gun and 88-mm fire on the advancing Americans. In the ensuing fighting, enemy fire killed Capt. Cecil Butler, commanding officer of Company F. The forward observer called for a fire mission, and the first rounds to land fell on them. Before he could lift the errant barrage, Companies E and F experienced heavy casualties. Arthur Ridings, a member of F, later observed: "There is no way to determine the number of men killed or wounded by our artillery, however, it did play a major part in the disaster that day."[39]

Lt. John Long of Lancaster, Pennsylvania, was with Company H that day and remembered that the forward observer realized the rounds were falling short but was unable to adjust it; the FDC apparently thought that Germans were making the request. When the enemy initially opened fire, the forward observer made an immediate request for a fire mission, but the first rounds fell short. He then asked for the fire to be lifted. Long noted that next: "the Battalion Commander, thinking this was a request by the enemy, asked him to authenticate, and because he didn't know how (I'm not sure a seasoned officer would know how), the Battalion Commanding Officer would not lift the artillery. I'm sure the Germans heard this transmission, so they continued the bombardment all night." He recalled that the forward observer was severely wounded in this action and probably did not survive.[40]

Long's comments are insightful and raise questions concerning the true source of the "friendly fire" that fell on the two rifle companies that day. If the Germans had intercepted the initial request for a fire support, could they possibly have fired the first rounds believed to be short and then continued their barrage uninterrupted? It is an interesting question, but one that can never be answered with certainty. Like the Japanese, the Germans took advantage of every opportunity to try to convince American soldiers that their own artillery was falling on them.

This incident illustrates the tremendous burden of responsibility forward observers had for bringing their training into proper use. If one accepts as fact that this was friendly fire, the observer was doubly damned, first, for using the wrong coordinates and, second, for not being able to prove the authenticity of his broadcast. Even if one accepts the premise that it was actually enemy fire falling upon the men, then the forward observer is still to blame, again for his inability to "authenticate" his broadcast. Granting his request to stop all U.S. fire would have made it obvious that any subsequent fire falling on them was German. The 345th Infantry Regiment's unit history for December 1944 describes this incident on the seventeenth, listing Company E's casualties at 49 percent.[41]

The 11th Armored Division experienced an incident very similar in nature but apparently an enemy trick. When a patrol from the 11th came under intense German small-arms fire, it quickly radioed for artillery support. Because the enemy monitored the transmission, as soon as the battalion reported that help was "on the way, German artillery promptly opened fire on the patrol. Mistaking the enemy artillery fire for American shells falling short, the patrol called immediately for a 'cease fire,' saving the Germans from a heavy artillery barrage."[42]

The parity that existed between American and German artillery did not mean that the latter was exempt from making mistakes. One would think that after five years of experience, German infantry-artillery coordination by 1944 would be nearly flawless. Yet even at that date, the Wehrmacht still experienced problems with coordination, communications, and even friendly fire. The same morning the 3rd Battalion, 345th Infantry was fighting near Medelsheim, the German 48th Grenadier Regiment, commanded by Lt. Col. Wilhelm Osterhold, was approaching Belgium through the Schnee Eifel.

Osterhold had two battalions of Volksartillerie for support. His forward observer was linked to a radioman who then relayed all requests for fire missions to the batteries; they were connected by a single strand of telephone wire. As soldiers of Osterhold's first battalion advanced, they encountered a series of American mines and trip flares. Soon the troops were cutting every wire they could find. In their zeal to avoid setting off any explosions, the Landsers inadvertently cut the telephone wire that was their sole means of communicating with their supporting artillery. Almost immediately, German shells began landing all around the colonel and his men. With no way to lift the barrage, their only hope was to keep advancing. Friendly fire killed or wounded 60 percent of the 48th Grenadiers that morning.[43]

The four-hour delay that Lieutenant Allee encountered when he requested a fire mission on observed targets may have resulted from corps command's inability to immediately account for the location of sectional boundaries due to the recent realignment of American forces in that sector. The 87th Division was fighting a sophisticated enemy who understood how to fight a modern war. The German army became adept at locating and using these crossover points to their advantage. In a postwar interview with *Field Artillery Journal*, German general named Thoholte claimed that "almost invariably German observers could plot the divisional sector lines upon the terrain merely by making notes of gaps between zones of fire of division artilleries. At times, German tanks and infantry made use of this information and aimed their attack at a sector line, knowing that the zone would not be covered by either division."[44] This gave them the ability to penetrate and exploit small, unprotected openings in the American defenses. Perhaps this was the tactic they used at Medelsheim.

The 87th Division's initial experience revealed how sophisticated the German army was in their practice of combined-arms doctrine, especially on the defensive. The Wehrmacht used snipers and forward observers effectively to hinder the Allied advance and, at this point in the war, made more effective tactical use of its armor. The Golden Acorn Division's first taste of real combat pitted it against an adversary that already had five years of wartime experience.

The Medelsheim incident also demonstrated how critical the forward observer's role was to the safety of infantrymen and their tactical success. These men were the eyes of the division on the ground as it moved forward and represented its vital link—the controlling link—between infantry and artillery. The events of December 16 made it clear that no matter how good the controllers of supporting fires were, command could still pose a problem. Lieutenant McGhee was able to stop the grazing machine-gun fire and drive away the tanks to their front by directing a responsive time-on-target on Medelsheim. But the long frustrating delay that Lieutenant Allee experienced obtaining permission to fire on German artillery he could plainly see was unwarranted and resulted in unnecessary American casualties.

Some might point to Medelsheim as an example of the criticism Michael C. C. Addams and others have made that "Americans employed too much top-down management, meddled in issues best left to subordinates, and demanded obedience to orthodox rules and doctrine that robbed the front-line soldiers of initiative. Too often, [an observer] had to ask permission to call in fire or exploit a situation in his immediate sector."[45] Martin van Creveld, who made a comparative study of U.S. and German fighting power in World War II, concluded that the Germans operated under less rigid regulations, which gave them an overall advantage in fighting efficiency.[46]

With regard to communications, there was a much closer technological parity between the Germans and the Americans than existed in the Pacific theater. The Wehrmacht, like the U.S. armed forces, had good, reliable telephones, but they experienced the same problems with severed lines that the Americans did. Using their radios, the Germans could listen to Allied transmissions and mimic them to try to countermand firing orders and create confusion. Because of the extreme humidity on the islands of the Pacific, the American experience with radios there was less satisfactory than it was in Europe. Yet field radios of the early 1940s were never entirely reliable anywhere. The malfunction of Lieutenant McGhee's set on December 16 might have led to even heavier casualties for Company L had he not on his own initiative gone forward to the top of the ridge under heavy enemy fire to use the company commander's radio to stay in contact with his field-artillery battalion. From there McGhee was able to put his artillery training to good use in directing fire. For whatever reason, two days after the Battle of Medelsheim, the 87th Division artillery communications officer found himself relieved of his duties and replaced by Capt. Robert Magee of the 336th Field Artillery Battalion.[47]

General Patton pulled the Third Army out of the Saar Valley shortly before Christmas and moved it to Belgium to help counter the German attack in the Ardennes. The soldiers of the 87th took a bitterly cold ride north in open trucks. Shortly afterward, the German army quickly regained control of all the border villages and locations in eastern Lorraine that the Third Army had taken. Another six to eight weeks would pass before the Seventh Army would expel the Wehrmacht from that area permanently.

| # Forward Observers in the Ardennes

The Bulge

American forward observers on the ground played a critical role during in defeating the German Ardennes offensive, commonly known as the Battle of the Bulge. For the first few days of a campaign that lasted for about six weeks, the Americans used their combined-arms tactics primarily on the defensive, while the weather kept most of their warplanes grounded. After Patton's Third Army entered the fray, U.S. military forces gradually went on the offensive. Artillery contributed heavily to the victory in the Ardennes, and observers on the ground had a key role to play in providing critically needed fire support even while frequently fighting in the same manner as the combat infantrymen they supported.

The Nazis purposely selected a target date when bad weather was imminent. As a result, the artillery pilots, who could quickly spot targets and direct rapid counterbattery fire, would not take an active part in the initial stages of countering the offensive. Nonetheless, American gunnery was so intense early on that experienced German artillery officers were convinced their adversaries had a ten-to-one advantage in guns and ammunition.[1] Although too high, that ratio gives some indication of how much artillery the Americans appeared to be firing. While the Americans had a huge advantage in mobility and ammunition, the near tactical parity that existed between German and American artillery was evident throughout the entire campaign. On the first day of the attack, German artillery played a key role in helping punch gaps in the frontline defensive positions.[2] The work of forward observers on the ground became crucial since the majority of fire was observed. American artillery depended upon them to help stall the enemy offensive. After the weather cleared, the Allies had both ground and aerial observation again.

Public perception of the Battle of the Bulge, molded by Hollywood, holds that the campaign, which began on December 16, 1944, was over within a week to ten days. Although the Wehrmacht had reached its western-most point of penetration into Belgium in about a week, the final restoration of both armies' lines as they were prior to the offensive took about six weeks, or until the end of January 1945. Over that period of time, the enemy bitterly contested every mile of the way, demonstrating repeatedly an exceptional skill in mobile defense, defense in depth, and counterattack. At least one German armored unit that took part had long experience on the Eastern Front.[3] The 87th Division's role, then, was to push the enemy gradually back to the German border. The heavily wooded terrain of the Ardennes could blunt the effect of artillery fire in some instances. Nevertheless, forward observers played a very active part in a variety of circumstances that often blurred the distinctions between combat arms and resulted in many casualties among their numbers.

Not only did bad weather ground the planes, but the initial German barrage also severely disrupted the American communications network. In some cases U.S. batteries fired directly on attacking German forces before being overrun. But despite these setbacks, American gunners were able to delay the enemy advance to some extent.[4] The experience of the 2nd Infantry Division in the early stages of the campaign demonstrated that.

In the 99th Division's sector, a road ran south through the twin villages of Krinkelt-Rocherath, intersecting a main east–west road at Bullingen. The paved highway headed west to Malmedy by way of Bullingen and Butgenbach, providing suitable access for German armor to reach Antwerp.[5] A stubborn American defense of the twin villages and road junctions in the surrounding area by elements of the 99th and 2nd Divisions gave the Allies enough time to regroup near Elsenborn.

On December 17 Lt. Charles Stockell, a forward observer from the 37th Field Artillery Battalion, was with elements of the

2nd Division at Hunningen when the Germans launched an attack toward the village following massive shelling. Stockell dashed forward and on reaching the church quickly climbed a series of fragile ladders to reach the steeple. From there he brought down a deadly barrage upon advancing German troops. The enemy eventually withdrew but made six consecutive attempts to dislodge the Americans during the afternoon and early evening. Stockell, however, stayed in the church tower most of the day while German artillery and tanks used direct fire to hit the steeple at least ten times. Finally, he and his radio operator raced down the steps. By the time they reached the ground, enemy shells had completely destroyed their observation post.[6]

After General Patton assured Gen. Dwight D. Eisenhower that his Third Army could help stop the German advance into Belgium, Eisenhower pulled Patton's command out of the Saar Valley, and by December 30 the 87th Division was in line near Libramont approaching Bastogne, about thirty miles to the northeast. In the fluid battlefield conditions of the Ardennes, forward observers often made initial and sometimes unintended contact with the enemy.

On January 4, 1945, Tech 4 Ray J. Jemc from Chicago, Illinois, of Battery A, 336th Field Artillery Battalion was already on his third forward-observation mission, this one led by Capt. Thomas H. Choate and joined by Sgt. Joseph M. Benicky, Jr., providing artillery support for the 2nd Battalion, 346th Infantry operating in the vicinity of Vesqueville. After moving through some woods, they came to a road and spotted an armored car heading toward them, an American vehicle the Germans had captured during the early stages of the Ardennes offensive. An exchange of fire between the Americans and the Germans soon broke out. The car came to a quick halt about seventy yards away. Captain Choate called for artillery fire, but the first rounds to arrive almost hit the observation party. The armored car was soon knocked out, however the three Germans had already abandoned it and fled into the woods.[7]

The area near Vesqueville was a hotbed of enemy activity that day. Lt. James McGhee felt certain that the Germans had an observer in the church steeple in the little town.[8] If the Americans had held the village, he probably would have used the same vantage point. PFC Robert C. Reed of Durham, New Hampshire, was in an antitank platoon of the Headquarters Company, 3rd Battalion, 345th Infantry.

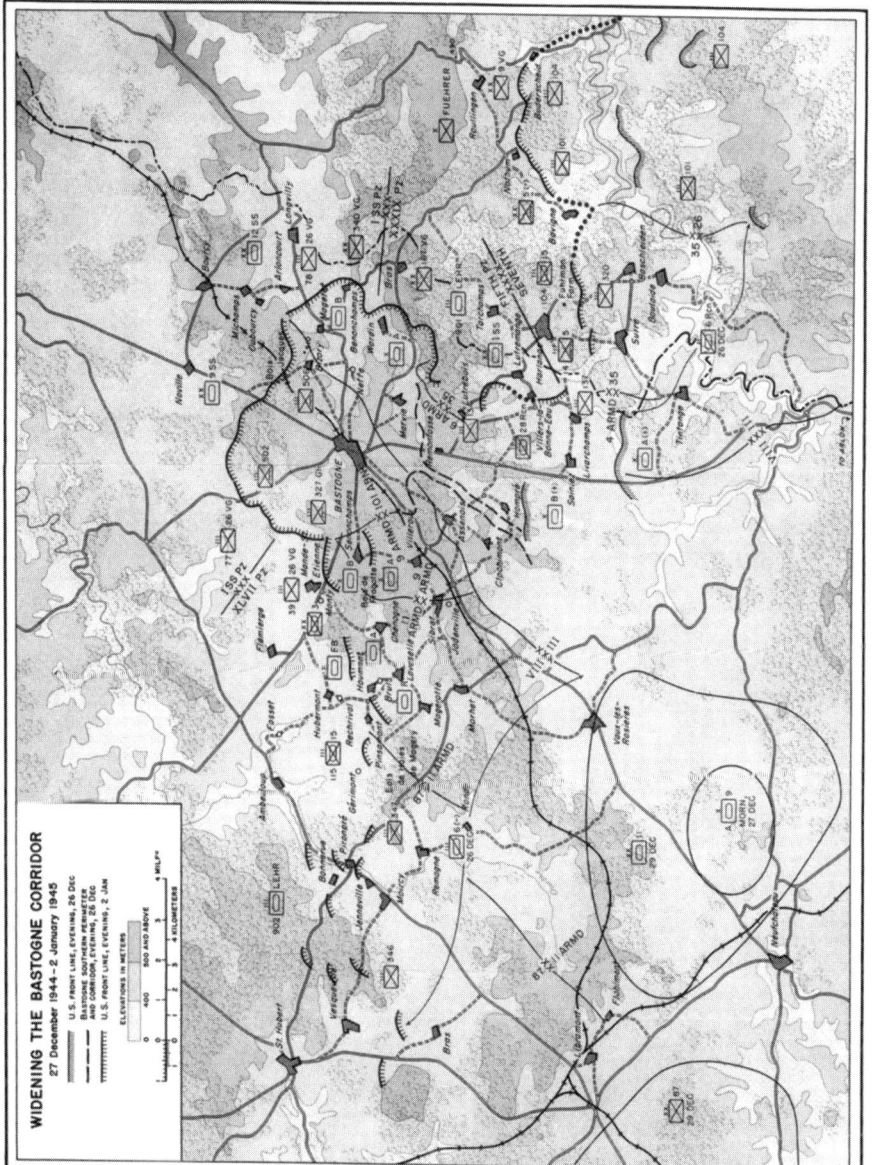

The 87th Division's position in Belgium From Hugh M. Cole, *The United States Army in World War II: The European Theater of Operations: The Ardennes: The Battle of the Bulge* (Washington, D.C.: Center of Military History, 2007).

Over the crest at Vesqueville, Belgium. The church steeple is visible to the right of the white gable. Photo by author.

His squad had just left Bras Haut and had repositioned its 57-mm gun on the edge of a woods overlooking Vesqueville. Reed remembered that his crew received the order to fire at a steeple supposedly holding a German observation party: "Artillery fire was being directed at one of the rifle companies preventing its further advance. We may or may not have hit our target, but disclosed our position. Consequently, we ourselves were shelled and two men died."[9]

After the 87th Infantry took Vesqueville, McGhee and his party returned to get their jeep and drove there. On the way they took a shortcut and found themselves stuck on top of a snowdrift. Getting out of the vehicle to shovel it free, they discovered that they had stopped in a minefield. With snow suspending the body of the jeep, the right-front wheel was about ten inches above the ground. McGhee recalled his party found a mine a few inches below it. They carefully had to remove several mines before they could shovel out enough snow to extricate their jeep. Afterward, they abandoned their attempted shortcut and took the long way back.[10]

At the Battery A reunion of 2008 in North Olmstead, Ohio, Rowland J. Coonradt, a veteran of the 334th Field Artillery, recalled a similar experience. Serving as a member of a forward-observer party, he was helping his lieutenant adjust fire while lying in the snow and noticed that he kept sliding around on something. When they were done with their fire mission, Coonradt discovered that he had been on top of a Teller mine.[11]

Other members of the 334th Field Artillery Battalion were not as fortunate in such encounters. On March 27 Maj. Franklin C. Seiler and Capt. Joseph G. Mobley were killed when their jeep struck a mine near Buchelborn, Germany, as they performed reconnaissance duty. Cpl. Rayford Willis and Pvt. John Perkinson, also in the vehicle, were seriously injured.[12]

In support of the 87th Division, VIII Corps artillery fired a heavy concentration upon the positions held by the Panzer Lehr. But the Germans had carefully concealed their tanks from view on the road by placing them behind piles of lumber surrounding a sawmill.[13] They not only held but also stopped a strong infantry-tank assault and drove the Americans back.[14] After the Germans counterattacked elements of the 347th Infantry near Bonnerue on the morning of January 6, the 1st Battalion, 345th Infantry relieved it. Everett "Bob" Criss of Company C remembered it as a cold, dark day with lots of snow on the ground. Casualties had reduced his unit to roughly sixty men, the second platoon to about the size of a rifle squad. His squad took up a defensive position inside a two-story house, with an attic and heavy stone walls, somewhere near the middle of the village. He later recalled: "My sergeant assigned me to guard duty in the attic to observe the hillside. When three German tanks appeared side by side, perhaps 1,000 to 1,500 feet away, coming over the brow of the hill, I ran downstairs to sound the alarm and then went back to my post in the attic."[15]

Two of the German tanks entered the village. When an American bazooka team disabled the first with a bazooka round, its crew evacuated. Then the second tank opened fire on the house with the platoon inside. Criss was severely injured, his right leg nearly blown off below the knee, and he passed out. When he awakened in total darkness, he was lying on his back in the cellar. Criss remembered: "I could hear an artillery FO calling on his radio to direct fire on top of us to drive away the tanks. I was in and out of consciousness most of the night. I was awakened when someone stepped or tripped over my right leg. The pain was excruciating. Someone else told me to be quiet because the Germans were all over the place."[16]

Criss heard Lt. Robert T. Booth of Plattsburgh, New York, the forward observer from Battery B, 912th Field Artillery, on the radio directing artillery fire that night. After the 345th relieved the 347th on January 6, Booth and his forward-observation party stayed to support Company C and the remainder of the 1st Battalion. The lieutenant used the same house that Criss and his squad occupied because its attic provided a good observation post.[17]

Criss later concluded that the forward observer helped save his life. "Because of the artillery fire they [the German tanks] withdrew during the night. Before dawn the next morning, a medic jeep drove into the field across the road from the house we were in." With the enemy tanks gone, the medics were able to evacuate him and another severely wounded soldier. By the time Criss reached the battalion aid station, it had been at least fourteen hours since his wounding, and he was sure that without receiving proper treatment, he would not have lasted much longer.[18]

The Germans were determined to recapture Bonnerue since it sat along an improved highway running from St. Hubert to the west to Magerotte and Bastogne, around six miles to the east. The German Seventh Army received orders to hold the Americans south of that road in order to maintain supply lines to their units attacking twelve miles farther west. The terrain surrounding the village gave the advantage to its defenders.[19] The history of the 345th Infantry described Bonnerue as a "vital road junction town on their supply line to St. Hubert."[20]

Heavy fighting continued around the village for the next seven or eight days. Artillery played a key role for both sides. Its near-even application resulted in a temporary deadlock. American artillery helped the defenders hold out, while German guns drove back repeated attempts by the 345th Infantry to advance through the woods to the north of the village.[21] Booth later observed: "Artillery in those days helped each side to gain and to lose the village several times. The net result was a fierce and bloody stalemate. Neither side could 'win' in the face of [the] artillery fire of the other. But use of the highway was denied to the Germans: our objective."

But this denial was only temporary. Late on the afternoon of January 7 following artillery preparation, German infantrymen, accompanied by two tanks, attacked again. A round from a bazooka destroyed one tank and an antitank gun the other. Booth recalled: "I directed artillery fire on them and on us all thru that. We were in a solid building and partly below ground so I drew a 'box' of fire around our position. My people supported me with heartwarming vigor. The 912th fired 614 rounds on the fracas—about 50+ rounds per gun."[22]

Early the next morning, following another artillery barrage, the Germans attacked again. Lieutenant Booth drew down the "box" of surrounding artillery fire so tightly that he set the house that was his position on fire. At that point his radio batteries went dead, and with them all hope of receiving any additional artillery support. With the building they occupied on fire, no communications with their own lines, and completely out of ammunition, surrender appeared to be their only rational choice. As a group they decided to capitulate. Describing the momentary fear they all felt that the Germans would open fire when they appeared before them, Booth later said: "Yet there was a strange feeling like that between two teams who have battled to a 14–13 score. Although emotions remained high, the game was over."[23]

The Germans took Booth and his comrades, SSgt. Stephen Z. Cieslak and Tech 4 John M. Watson, as prisoners, along with the members of Company C. Cieslak and Watson had helped Booth maintain radio contact during the fighting.[24] Fortunately, the Americans survived their captivity.

By January 1945 it was not unusual for noncommissioned officers to call and direct artillery fire. Many of them eventually received commissions as a result of the leadership qualities they displayed. In fact, the majority of battlefield commissions received in the field artillery came from the enlisted men of the forward-observer parties.[25]

The first ten days of the Battle of the Bulge took an unusually high toll of forward observers, the Americans fighting on the defensive almost continually, with little or no relief.[26] The 87th Division was still relatively new to combat when it arrived in Belgium. Yet as its artillery-battalion histories and other sources indicate, the unit had already lost a number of forward-observer personnel killed or wounded in the Saar Valley phase of the Lorraine Campaign.[27]

On January 4 Tech 5 Calvin H. Buchanan, Battery A, 912th Field Artillery, was serving as a forward observer near Pironpre. At dawn he was at a post in front of the infantry and about two hundred yards from a German position in the woods when he spotted two tanks. Despite enemy fire that concentrated on his position, he adjusted artillery on the German position so accurately that the tanks withdrew. For his quick and decisive actions, which averted a likely German counterattack, Buchanan received the Bronze Star.[28]

That same day another member of the same battalion, SSgt. Elliott S. Greenberg of Battery B, was in charge of an observation detail with a rifle company. As they began initial assaults upon German positions, the Americans encountered heavy small-arms and direct fire from self-propelled howitzers. Despite being wounded in the face by a shell fragment, Greenberg remained with the detail and continued to perform his duties as forward observer. Three hours later, after he had had received a second serious wound, he agreed to be evacuated. Greenberg also received the Bronze Star.[29]

Historians generally credit the ability of American enlisted men and noncommissioned officers to seize the initiative and fight small-unit actions in the absence of their superior officers as the deciding factor in the Battle of the Bulge. As Brig. Gen. Hal C. Pattison noted: "The story of the Ardennes then is the story of the American fighting man and the manner in which he fought a myriad of small defensive battles. . . . It is the story of squads, platoons, companies, and even conglomerate scratch groups that fought with courage, with fortitude, and with sheer obstinacy, often without information or communications or the knowledge of the whereabouts of friends."[30]

Although the Germans made frequent counterattacks, the 87th Division was generally fighting an offensive action in the Ardennes. Throughout the battles, its field-grade officers, noncommissioned officers, and enlisted men also displayed the same individual initiative. The quick actions of Cpl. Richard G. Hildebrand and PFC Clyde L. Jackson, both members of Battery A, 336th Field Artillery, provide one such example. On January 8 a German tank suddenly appeared, threatening the infantry that the two enlisted artillerymen

accompanied. Working together in the absence of their superior officer, they directed artillery fire against the tank, forcing it to withdraw,
although their training had not included control of artillery fire.[31]

That same day, Hitler ordered Gen. Sepp Dietrich's Sixth Panzer Army to begin to withdraw to areas northeast of St. Vith and
east of Wilz.[32] The 347th Infantry's regimental history states that on
January 10, "Bonnerue and Pironpre were taken, the Haies de Tillet
Woods were finally cleared of enemy forces and contact was established with the 345th Infantry on the right flank."[33] On that day
the forward-observer party of Captain Choate, Sergeant Benicky,
and Private Jemc was accompanying elements of the 346th Infantry
Regiment as it approached Tillet. After coming under heavy fire,
two units of infantry lost their platoon leaders. Benicky quickly reorganized and led the riflemen, acting to restore order by reassuring
the men. By his quick actions, he bolstered the men's confidence,
enabling the entire company to ward off a German counterattack.[34]

As the Third Army gained control of the area around Bastogne
and the 87th Division sealed of the roads west of the city, the remnants of German forces in Belgium began shifting farther east
toward St. Vith. By January 14 the 346th Infantry began moving to
the vicinity of Aspelt, Luxembourg, followed by the 347th the next
day.[35] By the seventeenth the 345th Infantry had shifted south to
Echternacht, a small Belgian town in the southwest corner of the
Bulge.[36]

Although fighting did not entirely cease during the division's stay
in Luxembourg, its only contact with the enemy came by patrolling.
On January 24 the 76th Division relieved the 87th in Luxembourg.
Two days later all units of the 87th were in the general vicinity of
Limerle and Hautbellain, Belgium, with the exception of the 335th
Field Artillery. Even at this date German lines continued to lie west
of where they had been prior to December 16, and the battle to
eliminate their gains continued. The ordinary German soldier now
had a new motive to fight—to keep the enemy from invading his
homeland. He bitterly contested every mile.

In the early hours of January 28, the 3rd Battalion, 345th Infantry moved to positions east of St. Vith to relieve the 7th Armored
Division. By January 29, regimental headquarters had also moved
to St. Vith, and the 345th was again in position to attack. The division now lay on the extreme eastern border of Belgium.

On January 30 the battalion attacked and captured Hodgen, the next day resuming the attack and capturing the border villages of Amelscheid, Heuem, and Atzerath through waist-deep snow.[37] Heuem did not fall easily, and the attack required exceptionally close coordination of artillery fire with infantry. During the day, Company A entered the village twice, but both times a stubborn enemy drove it out. Finally, by midevening both Companies A and C were firmly established in its streets. After capturing Setz, Lt. Col. Robert B. Moran's 3rd Battalion moved toward Heuem to assist the 1st Battalion, but enemy assault guns and artillery positioned on nearby hills hindered this advance. Moran then called for artillery support, and the 334th Field Artillery Battalion fired the mission, relaying the corrections from the forward observer to Colonel Sugg at the regimental command post and on to the FDC, enabling the battalion to continue its advance.[38]

"On the morning of the 30, the 3rd Battalion [346th Infantry] jumped off toward Andler but were unable to cross the [Our] River into town. At 0400 [on] the 31st, the 1st Battalion passed through the 3rd Battalion and . . . after a day long and all night fight[,] captured Andler at 0530, 1 February. The 2nd Battalion continued to hold and secure positions in the vicinity of Schonberg."[39] The 87th Division was now about to reenter Germany.

After the Battle of the Bulge had ended, the commander of the First Army, Lt. Gen. Courtney Hodges, indicated that although the German defeat was the result of all First Army arms and services, bad weather and the rugged terrain had put the fighting capabilities of the infantry, armor, and air at a serious disadvantage. He further observed: "Through all however—day and night, good weather and bad—the flexibility and power of our modern artillery was applied unceasingly. A lesson then from the Battle of the Bulge—Artillery constitutes a most formidable striking power continuously available to any commander of combined arms for application wide and deep over the battle area."[40]

Of course, all combat arms contributed to the successful delaying action that took place before Christmas 1944, allowing U.S. forces to regroup. After the Germans captured the twin villages of Krinkelt-Rocherath, American infantry first took the ground, then called for armor, tank destroyers, and artillery support. American artillery, controlled by observation teams working in close coordination with infantry, was essential in applying fires on the enemy. Lt. Col. William D. McKinley of the 38th Infantry made heavy use of it in his battalion's defense of the Lausdell crossroads near the twin villages, halting a number of German panzers. Emphasizing the degree of combat leverage that effective artillery support could provide, McKinley later wrote: "It was the artillery that did the job. On three different occasions artillery support when and where it was vitally needed saved my battalion from decimation and the last time from complete destruction."[41] Summarizing that same series of battles in the early stages of the Ardennes campaign, the *Combat History of the Second Infantry Division* notes that "massed artillery fire from the four organic battalions [of the Second Division] was used to greatest advantage attested by staggering losses to the enemy. In large measure, it helped to turn the tide of the German thrust."[42]

The ability then to provide continuous, close artillery support through the use of forward observers was a key factor in the successful use of American firepower in the Ardennes. From their elevated vantage point, aerial observers covered more ground, generally saw more, and spotted some targets that observers on the ground would never see. But this was contingent upon the ability of the planes to take to the air and the conditions of visibility existing at any given time. Poor visibility and bad weather conditions prevailing throughout various phases of the Ardennes Campaign negated this advantage, making the work done by forward observers on the ground even more critical to slowing the German advance and to the eventual American success.[43]

The Battle of the Bulge was the largest and single-most costly campaign in U.S. military history. Hugh Cole has noted, "the attack of twenty-nine German divisions and brigades destroyed one American infantry division as a unit [the 106th], badly crippled two infantry divisions, and cut one armored combat command to pieces."[44] His account of American casualties in the Ardennes only covers through January 2, which he acknowledges was probably incomplete. Through that date, 4,138 Americans reportedly died in battle, 20,231 suffered wounds in action, and 16,946 were listed as missing. The army's final report of combat deaths lists casualties incurred in the Ardennes and Alsatian campaigns together for the period December 16, 1944–January 25, 1945. It lists 12,359 killed in battle among all combat divisions; another 53,333 as wounded in action, of whom 1,958 subsequently died from their wounds; and another 20,582 as captured, of whom 482 died while interned and 2,529 remain missing in action.[45]

On February 2, 1945, the 87th Reconnaissance Troop was ordered to capture the German town of Roth, and the next day the 87th Division began entering Germany for the second time since mid-December 1944. Two days later the 345th Infantry occupied the area west of Kobscheid.[46] In this tiny rural village in the Schnee Eifel region of western Germany, the Landsers of the 295th Regiment, 18th Volks-Grenadier Division had overrun Troop A, 14th Cavalry attached to the 106th Division seven weeks earlier.[47] Now the German gains in Belgium had been completely eliminated. With little or no rest, the 87th Division's next mission would be to penetrate the Siegfried Line—again, for the second time. While the heavily wooded terrain would, at times, hinder the ability of forward observers to conduct fire, they would continue to contribute in a variety of ways to the Allied victory.

The ferocious struggle in the Ardennes Forest revealed that the Germans were still very capable of mounting an effective offense, utilizing their experience with combined-arms tactics against a thinly held American line. Fortunately, the United States reacted quickly, committing its reserves much more rapidly than the Germans had anticipated.[48]

The actions of forward observers in this campaign demonstrate the effectiveness of American combined-arms doctrine, first in restraining the Bulge, then in reducing it. These artillerymen had to contend not only with enemy infantry but also with German armor, often assisted by deadly artillery fire. Although fighting the Japanese had its own set of unique dangers, forward observers in the Pacific theater did not routinely face such a well-coordinated, three-pronged offensive punch.

The experience of the 87th Division reemphasized that forward observers responsible for controlling artillery fire could only do so effectively when situated at the infantry's most forward and vulnerable positions. In all actions they became so fully integrated with combat infantrymen that their designation as artillerymen sometimes became meaningless. Also, the distinctions between infantry and artillery became blurred as officers of either combat arm simply became leaders during battle and did what needed to be done.

One lesson repeatedly drawn from the individual stories of battle in the winter of 1944–45 was that effective practice of combined-arms warfare forced the complete integration of fighting men regardless of training or rank. Repeatedly, enlisted men and officers alike, in the absence of their superiors, stepped in to fill leadership roles to get a job done. The actions of forward observers fighting in the Ardennes also proved true General Pershing's words from years earlier, that the successful execution of open warfare requires "individual and group initiative, resourcefulness, and tactical judgment," even if his concept of how that could be achieved was flawed.[49]

| # Through the Wall and across the Rhine

By the time the 87th Division had punched through the Siegfried Line and crossed the Rhine River, its capacity for executing combined-arms warfare had fully matured. After breaching the Westwall (Germany's name for the Siegfried Line), capturing the ancient city of Koblenz, and crossing the last natural barrier to Germany's interior, the division's daily progress accelerated beyond anything it had previously known. Forward-observer parties accompanying the various task forces continued to contribute to victory in ways that transcended the boundaries of their designated specialties.

The effort to crack the Siegfried Line meant not only using forward observers at the front to direct artillery but also infantrymen calling on mortars and armor to apply the firepower necessary to break through the line. The many accounts of inspirational service by forward-observer teams confirm two important lessons learned. First, in the final months of the war in Europe, the distinction between infantry and artillery became blurred as officers shouldered responsibility, when necessary, to save the men around them and complete their mission. Also, the distinction between officers and enlisted men broke down as the final drive to victory necessitated using artillerymen of all ranks to direct and adjust fire on enemy targets.

On February 2 the 345th Infantry established its regimental headquarters between Wischeid and Auw, just east of the Belgian

Sixty years after the war, decaying "dragon's teeth" tank obstacles still lie in a field near the German-Belgian border. Photo by author.

border. Two days later Major General Culin ordered the 3rd Battalion to Kobscheid and the 2nd to Auw in preparation for the regiment's upcoming assault on the Siegfried Line; portions of the Westwall were not much more than a mile east of Kobscheid. On the morning of February 5, the regiment received orders to attack the line in two places. Less than two miles due east of Kobscheid was an important crossroad the Germans used to move troops and artillery. The second objective was another intersection about two miles south of the first one. Colonel Sugg assigned the first crossroad to the 3rd Battalion and the second to the 2nd Battalion. Both intersections were linked by the Schneifel Hohenweg, an improved road running northeast–southwest and protected on both sides by a series of fortified bunkers running parallel to it—the first fortifications of the Westwall in this sector. The 2nd Battalion jumped off

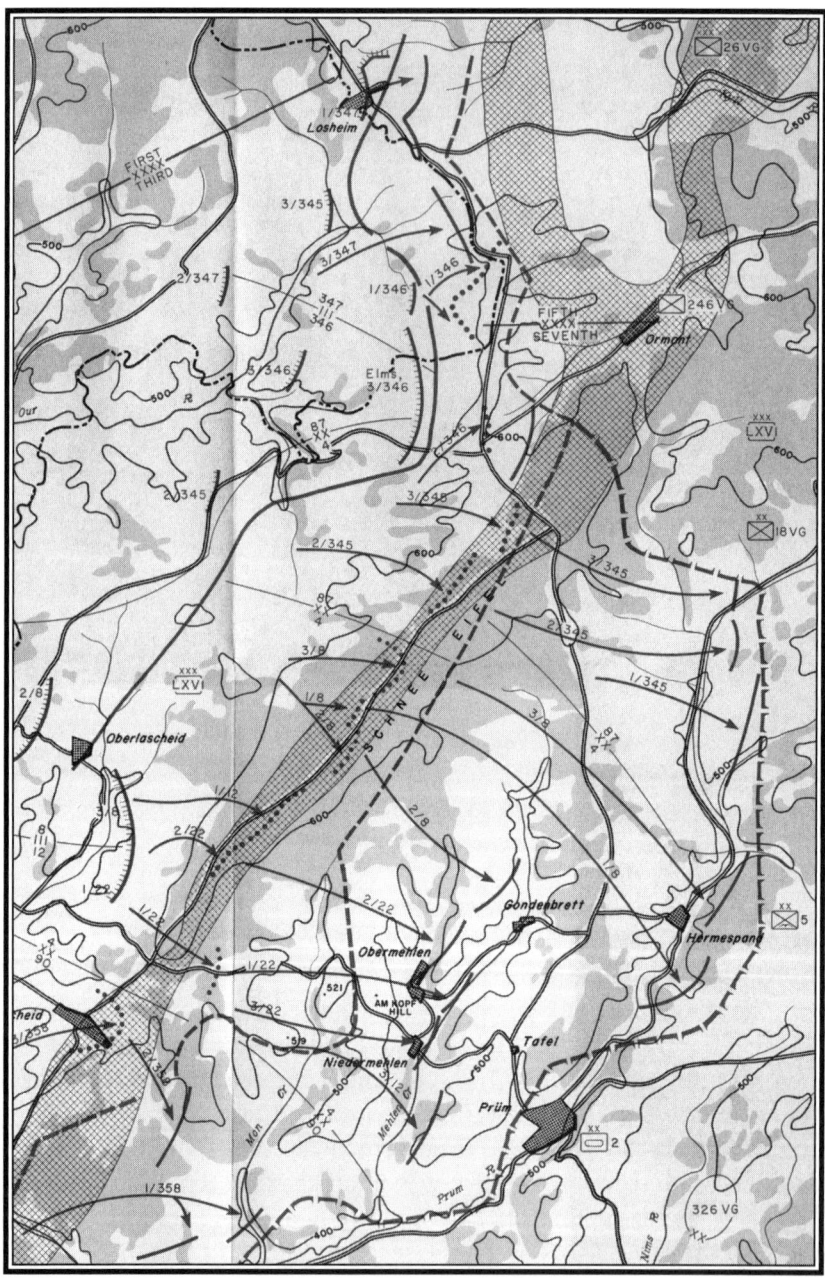

The 87th Division's attack on the Siegfried Line through the Schnee Eifel, February 1945. From Charles B. MacDonald, *The Last Offensive* (Washington, D.C.: Office of the Chief of Military History, U.S. Army, 1973), map 4, "The Drive on Prüm."

138

View of the Schnee Eifel from Kobscheid. A line of Westwall bunkers extended below the horizon, left to right. Photo by author.

about 8:00 that evening and by 11:00 had passed through Kobscheid heading southeast.[1]

Lt. James McGhee and his two enlisted men accompanied Company F to provide forward observation for the 2nd Battalion. The infantry's advance was made on a very dark night under blackout conditions along a dirt road not much wider than a trail. A bulldozer followed the riflemen to clear a wider path so that vehicles could resupply the frontline troops. McGhee was with Company F's commander, Capt. Richard McCann of Columbus, Ohio, but the bulldozer driver became disoriented. With the captain's permission, McGhee then got on the bulldozer to guide the driver. After reaching the village, he jumped down and ran ahead to catch up to McCann.

The plan called for Company F to follow this narrow road southeast to a fork, where it was to take the left branch. But by the time McGhee had rejoined the company, Captain McCann had followed the wrong branch of the fork. Immediately the lieutenant told McCann that he was on the wrong road and heading into an area where a massive artillery preparation was scheduled to fall very shortly. The captain disagreed but said he would have the company remain in place while he went back to check. Recalling the incident, McGhee wrote: "I told him we did not have that much time before our artillery would start coming in on us. He then walked back . . . and after he had gone a short distance our shells started whistling in just in front of and almost on top of us. I took the responsibility of grabbing the platoon leaders and moving them back as quickly as possible toward the fork in the road."[2]

As this drama in miniature was unfolding, McGhee ran into another problem. One of the platoon leaders, a young lieutenant, was cowering in a ditch and weeping. When McGhee told him to gather his men and get moving, the officer replied, "I can't! I can't!" At this point McGhee slapped the man across the face and told him, "You have more to fear from me than you do from the enemy." Thus energized, the young lieutenant got on his feet and began to move his men back to safety.[3]

The training forward observers received included map reading and orienteering exercises, just as it did for infantry officers. It is understandable that anyone could become lost in an unfamiliar setting in total darkness, but it was also fortunate that Company F had an artillerist with them who was able to follow the plan. It was also lucky that McGhee took the initiative to order the men back to safety. While his treatment of the young lieutenant may appear to have been harsh, friendly fire wounded none of the troops.

The distance from Kobscheid to the crossroad objective was not much more than a mile, but it took all night to reach the outer portions of the Siegfried Line. The late-night artillery feint helped because the enemy did not see the battalion advancing in the early morning hours. The first German soldier McGhee saw was a lone sentry, whom the point man shot. After the company moved on, a firefight developed in a heavily wooded area. As this was taking place, Lieutenant McGhee and his two enlisted men, Sgt. Donald Walker and PFC Gayle Bricker, had been keeping pace with the infantry while crawling in a ditch. Just as the sun was rising on February 6, Walker tapped McGhee on the heel of his boot and pointed

An intact Westwall bunker in the 63rd Division's sector. Shortly after the war, the French demolished most of the bunkers in the Schnee Eifel area. Photo by author.

to a German pillbox they had passed in the predawn light. The two decided they would try to capture it. Bricker witnessed what happened next: As the two men neared the position, they approached it from opposite ends. Walker cut the phone lines. As the bunker's sentry came out, Walker motioned with a white phosphorus grenade for them all to come out. He did not even hold a gun on the German soldier.[4] Ten men came out, and Captain McCann dispatched a rifleman to take the prisoners back to Kobscheid.[5] The 87th Division's record of enemy activity for February 6 noted, "FO and Sgt captured pillbox and ten Germans."[6]

That same morning Lt. Lester T. Price, forward observer with the now badly battered 3rd Battalion, had been wounded, and Lieutenant McGhee and his section were sent to take over his assignment in the vicinity of the 3rd's objective to the north.[7] Just before dark McGhee and his group reached Company L's command post, where they spent the night.

On the morning of February 7, the 3rd Battalion, 345th Infantry continued its advance toward the crossroad objective when a heavy firefight erupted. After the crossroad was secured, a German prisoner disclosed that to proceed further would require crossing an extensive minefield and the reduction of several more pillboxes. Lieutenant McGhee volunteered to cross the minefield, using a German sergeant to lead the way. Accompanying McGhee were Sergeant Walker and Cpl. John M. Gazvoda of Detroit, Michigan. Both volunteered to go with him and an infantryman, Hallis C. Workman of Company I from North Canton, Ohio. In less than a week, young Workman would suffer mortal wounds.[8]

The German sergeant was reluctant to lead the way, but with McGhee's pistol stuck in his back, he agreed to cooperate. The occupants of the first bunker they encountered were members of the *Volksturm,* the conscripted national militia Hitler founded in October 1944 of German men between the ages of sixteen and sixty not already in the military. With a little oral persuasion, they readily surrendered. The troops in the second pillbox wanted to surrender, and when they saw the Americans coming, they rolled up their bedrolls and were sitting on them by the time McGhee and company arrived, ready to be led back as prisoners of war. The third bunker was flooded and empty.

The last bunker was the largest of the four. Somehow they had approached it undetected. McGhee took his white phosphorus grenade, pulled the pin and announced that he would throw it in unless all the occupants came out, all the while the German NCO pleading earnestly for his countrymen to surrender. Fortunately for the four Americans and their reluctant companion, someone inside stuck a white rag through the gun aperture, and twenty-two German soldiers, including three officers, soon came out.[9] Years later, after reading a brief account of this incident, Don Walker recalled that the Germans had come out of the pillbox without their weapons. McGhee told him to go in and cover the small detachment of prisoners sent in to retrieve them.[10]

The actions of McGhee and Walker were not unusual for artillerymen in their position. Their boldness and willingness to take the

The rear of a large bunker now used as a hunting lodge. This is in the vicinity of the last bunker Lieutenant McGhee's men captured. Photo by author.

initiative demonstrated once again the blurring of combat-arm distinctions in the field. Both McGhee and Walker received the Silver Star for their deeds.[11]

Leadership and bravery can be contagious qualities in combat, and the 87th Division's forward observers had many good examples among their infantry officers from which they could take inspiration. Capt. Howard J. Wall, Company L's commander, and Lt. Col. Robert Moran, commanding officer of the 3rd Battalion, 345th Infantry, were two such men. Unfortunately, Wall would not survive the war.

At midmorning on February 7, Company L assisted in the assault upon the 3rd Battalion's objective east of Roth.[12] PFC Augustus "Gus" Epple of Cape May Court House, New Jersey, was an ammunition bearer in a mortar squad in Company M and later wrote a description of Wall's actions he witnessed that day. As the battalion began its advance that morning, Epple and his section sergeant, Frank Kosub, were still with their mortar squad. The rifle companies preceded them along an uphill course beside a woodland screening them from the pillboxes. Suddenly they stopped, and shortly afterward, a heavy barrage from enemy artillery and mortars began falling on them. Epple wrote later that he was not hurt but was "mad as hell" that they dared shoot at him, adding: "Frank and I were then assigned to L Company and were instructed by the company commander, Capt. Wall, to stick with him and he would call for mortar fire. L Company was the lead rifle company and we went into the woods (instead of parallel to them) facing the pillboxes."[13]

After the 3rd Battalion started moving east along the highway, a German machine gun opened fire on it. Captain Wall ordered the two tank destroyers supporting his company to advance. As they moved up, he directed their fire on the fortification and put it out of action. Company L advanced until stopped again by a second pillbox.[14] By this time, Lieutenant Colonel Moran had joined Wall and the leading platoon of rifleman and called for fire. Epple wrote: "The gunner made the shot of his life, hit the gun slot and momentarily stunned the German squad inside. Moran grabbed the bazooka, ran around to the back entrance, and fired at the door. Capt. Wall followed close behind, tossed in a grenade, and 3 Germans came out coughing and bleeding. We had captured our first pillbox very late in the afternoon."[15]

Soon after the first bunker had been taken, Wall took a second by himself. "Directing two men to cover him, Capt. Wall worked his way forward alone until he reached the entrance of the bunker. Throwing in a hand grenade, he awaited the explosion, dashed inside a second later, and emptied his submachine gun at the occupants. His score—two killed, six prisoners taken, and another pillbox destroyed."

Later, after a concentration of enemy artillery fire left several men of the company wounded and lying in exposed positions, Captain Wall, acting alone, moved across open ground to reach two of the men and bring them back to a sheltered area where their wounds could be treated and they could be evacuated. A little later enemy machine-gun fire stopped Company L for the third time that day. Under fire, the captain ran to a building, where he was able to spot the position of the enemy emplacements. Then he ran back to his men and again brought up the tank destroyers. They destroyed one bunker with a direct hit and forced the occupants of the other to surrender.[16] Tragically, Captain Wall, who was from Troy, New York, was killed two days later.

These acts of courage displayed by Wall and Moran were inspirational to the combat infantrymen under their command as well as those medics, engineers, forward observers, and others serving in the front lines. Perhaps their brave deeds served to reinforce each other, exemplifying combined-arms tactics at their best. In any case Lieutenant Colonel Moran recognized the contributions made by his forward observers in combat during the breaching of the Siegfried Line. A few weeks later, when things had temporarily quieted down, he awarded the Combat Infantryman's Badge to Lieutenant McGhee and Sergeant Walker. Years later Walker wrote: "Div. Hdqtrs said that only bona fide infantry and engineer troops could wear them [the Combat Infantryman's Badge], so we only got a pleasant memory of the fact that one Infantry officer was aware that artillery FOs were in the war too."[17] Wall and Moran had the respect of all the men who served under their command.

The 345th Infantry's next objective was Olzheim, around one and a half miles to the south. By 2:00 P.M. on February 8, Company F was about a mile southwest of Olzheim, Company G was about seventeen hundred yards to the south, and the 1st Battalion swept around behind to reach the high ground overlooking town.[18] Edwin C. Pancoast, then a platoon sergeant with Company G, wrote that "what made the seizing of Olzheim so important was the telephone cable linking the German high command with the West Wall [that] ran through the center of town. Without telephone connections, the Germans would be forced to use radio communications, all of which traffic the US could readily decode."[19]

Pvt. Lane Barton of Vancouver, Washington, from Company G, participated in the attack on Olzheim. Later he described the extreme accuracy of the artillery support provided for his battalion that day after his squad had dug in to the side of a steep hill overlooking the valley. From their vantage point, they spotted a column of German tanks speeding along a road far below. The forward observer gave his first order for fire, and a short while later, the rounds went rustling by overhead. Barton recalled: "The FO's initial fire order put his ranging fire close to the middle of the column. He called; 'Fire for effect,' and three salvos of three rounds each burst in the middle of the tanks. One tank was hit directly. We could not believe our eyes. The FO acted as if the barrage was nothing out of the ordinary. I did not need further proof that our gunners were remarkably accurate."[20] Yet while the mechanical tolerances of the howitzers and munitions were very precise and the gunners who fired them well trained, it required the visual sensing and judgment of the forward observer to be able to direct and adjust fire with such accurate results.

The observer who destroyed the tanks with his accurate adjustments may have been either Lt. William T. Haun of Battery A or Lt. Joseph G. Turley of Battery B. The unit history of the 334th Field Artillery Battalion for February 1945 reports that these two men were both involved with massing fires on the towns of Olzheim and Neuendorf on February 8, indicating: "In one instance, Lt. Turley and Captain [James M.] Pollock, Liaison Officer, allowed the enemy to reinforce a garrison in Olzheim with several personnel-carrier loads of troops, then fired a TOT on the town. Enemy troops were disorganized & driven onto slopes east of town in scattered groups which were effectively fired upon by Lts. Haun & Turley."[21]

Company G, meanwhile, had met heavy enemy resistance. The company commander, Lt. Theodore Knusman, and the only other infantry officer present became casualties. Forward observer Lieutenant Turley was the sole remaining commissioned officer with the unit. Turley supervised the successful reorganization of the company until he was relieved.[22]

The 334th Field Artillery Battalion's forward observers were particularly busy that day. Late that afternoon Lt. Claude G. Wilson, Battery C's executive officer, serving as the forward observer, adjusted a bracket of fire on a German machine-gun nest. A white flag quickly appeared and twenty-two Germans surrendered.[23]

The 3rd Battalion, 345th Infantry lost one of its bravest officers in Olzheim on the morning of February 9. A German sniper shot and killed Company L's Captain Wall as he was directing the movement of his men in cleaning out the town. Howard Wall was known to all the men in the battalion as a courageous and fearless leader.[24] But the captain's death was not in vain. Ed Pancoast was correct; a German communications cable did run through Olzheim. Yet it appears unlikely that the 2nd Battalion's original objective was to capture the town for the purpose of disrupting German tele-phone communications simply because the cable's presence was not known at that time. By the ninth "all three Battalions were in the twin villages of Olzheim and Neuendorf, with the enemy watching every move from the high ground to the north and the east."

On the morning of February 10, Lt. Col. James B. Evans, 87th Division signal officer, made a surprise discovery as he studied the maps at headquarters. Evans noted that the foremost elements of the regiment had already overrun what was suspected to be the route of an underground cable connecting two key sites in the Sieg-fried defenses, Cologne and Bitburg. He and his men, accompanied by four members of the antitank company, set out for Olzheim and, after reaching an area south of the village, successfully located the cable. Using mine detectors to locate the wires, the men dug several deep trenches before striking "pay dirt." Despite a steady rain and harassing mortar, artillery, and small-arms fire, Evans's detachment cut the cable in two places and removed a large section.[25]

This seriously impaired the enemy's already weakened com-munications system in that sector. German general Richard Metz, senior artillery commander of the 5th Panzer Army, described the disadvantage this created for his forces fighting in the Schnee Eifel. By that stage of the war, it was nearly impossible for the German army to maintain central command of its infantry and artillery. Metz explained: "The reason for this was the fact that all telephone communications were not only interrupted for hours at a time by superior enemy artillery, mortar fire and aircraft bombing, but they had also been cut to pieces for long distances and could be repaired only by constructing anew, in lengths of kilometers, mostly at night." Without telephone service, German forces in this area had to rely almost solely upon radio communications.[26]

The 346th Infantry's regimental history reported: "For the next eighteen days the Division held defensive positions. On 26 Febru-ary 1945 the 3rd Battalion [346th Infantry[, was given the mission of breaching the formidable fortifications of the Siegfried Line near Ormont, Germany," a town along the Schneifel Hohenweg, per-haps six miles north of Olzheim. "The key terrain feature of enemy defense was Gold 'B' Hill, from which the enemy could effectively . . . observe artillery, mortar and nebelwerfer [literally, "fog-thrower," a high-explosive rocket] fire. . . . This hill also dominated the road net leading east from Ormont to the Kyll River. The hill contained fifteen twenty-man pillboxes. At 1330 on 3 March 1945 the battalion launched its attack . . . capturing the hill."[27]

Ray Jemc was with a forward-observer party involved in that action. Jemc recalled that he, Lieutenant Hollaman, and a new sergeant he referred to as "Fifth" were on their way to relieve Captain Choate's crew when they were attacked. The captain called for artillery fire and, along with infantry fire, stopped the attack. Choate's party had just cleared the area when German artillery opened up. During the ensuing barrage, a shell fragment struck Hollaman in the leg. Jemc wrote: "I pulled it out and gave it to the lieutenant and told him it did not look too bad. However, I did not tell him that it went in the back of his leg and came out the front of his leg. . . . Before he was evacuated, he gave me the maps of our area and told me what artillery fire to call for in case we were attacked again."[28]

By the last day of the month, the 345th Infantry was back in the line too, with the 3rd Battalion seizing the high ground east of Neuendorf. Tech 5 Don Welever of Wadsworth, Ohio, from Battery B, 334th Field Artillery, was serving as part of a forward-observer party that day. As the infantrymen his group accompanied began assaulting an enemy position, a German machine gun opened fire. Welever and his comrades took refuge in a series of vacant wooden bunkers. From these he quickly detected the origin of the enemy fire but also became aware of an American tank that had pulled up some distance behind his position. Seeing that the tank commander was taking aim and apparently about to fire upon the wooden bunkers that Welever and the other Americans were occupying, he quickly raced across one hundred yards of open terrain while under heavy enemy fire and jumped up on the tank, pounding on the hatch with his carbine to get the crew's attention. After he pointed out the location of the enemy machine gun, the tank quickly destroyed it.[29]

Although Welever's team may have been using a field telephone rather than a radio, it is unlikely that they would have been able to contact the tank crew by radio. Jonathan House notes that, even by this late stage in the war in Europe, "the radios issued to infantry, tank, and fighter aircraft units used incompatible frequencies, making communications among the arms impossible." Welever's brave actions prevented a friendly fire incident and probably saved the lives of the tank crew as well as those of many infantrymen.[30]

Over time, more commissioned officers designated as forward observers became casualties. The 334th Field Artillery history records, "since we lacked available officers to replace them, we had to call on enlisted personnel. Staff Sgt. [Rufus] Greening of Battery A, Staff Sgt. [Bennie] Morycz of Battery C and Tech 4 [Don] Walker took over the job."[31]

On February 28 the 1st and 2nd Battalions of the 345th Infantry set out to capture Neuenstein. Lieutenant McGhee was serving as the forward observer with Company E and had a new crew by this time. He recalled that when Sergeant Morycz became a forward observer, one of the crew members on his first mission became a casualty.[32]

Later McGhee remembered that the Germans repulsed an attack up a wooded hill before counterattacking. After both sides temporarily held the high ground, they soon moved apart so that the Germans held the east slope and the 345th Infantry the west. McGhee later recalled: "After things had been very quiet for a while, I attempted to make my way back to check on Morycz and his new crew (including Cpl. Robert J. Gehrt) only to be chased back by massive German machine gun fire. Before I could get to them I learned that Cpl. Gehrt had been hit by what was believed to be a 150 mm enemy shell."[33]

Gehrt normally served with Battery C as a truck driver, but on this occasion he volunteered to go forward to serve as a radio operator with a forward-observer party. Shortly after the battle began, he began communicating from a foxhole. He had just moved to a church tower when an enemy artillery shell made a direct hit on his location, killing him instantly.[34] Also wounded almost the same day while serving with another team was Corporal Gazvoda, who only weeks earlier had voluntarily accompanied McGhee across a minefield to the last of the bunkers captured that day.[35]

Meanwhile, the fight for Ormont continued. For their courageous actions on March 1, two members of Battery C, 912th Field Artillery serving with forward-observer parties received Bronze Stars. Tech 4 Donagh O'Hara and PFC Peter Hernandez aided in the destruction of an enemy roadblock that had brought the American advance in that sector to a complete halt. Both men worked to a position within 150 yards of the strongpoint. From there O'Hara operated the radio while Hernandez helped spot the bursts of

American fire. Both men continued this effort, sheltered only by a log while under intense mortar and small-arms fire, until the roadblock was destroyed.[36]

O'Hara later recalled that Hernandez had saved his life on more than one occasion. He also remembered that the artillerymen serving with forward-observer parties generally had a good rapport with the infantrymen with one exception. More than once while O'Hara was carrying a field radio, as he came under enemy fire, he quickly jumped into a foxhole occupied by a rifleman only to be told to "get the hell out!"[37] This was because the long antenna attached to the radio tipped off the enemy that here was an important target linking the carrier to either the artillery batteries or the higher echelons of command. Inevitably, the sight of an antenna almost always drew a heavy concentration of fire.

At 11:10 A.M. on March 1, three battalions of Division Artillery, plus the 687th Field Artillery fired a time-on-target mission on Gold "B" Hill, known to the GIs as "Gold Brick Hill."[38] Although Ormont fell that same day, the enemy still held the high ground on Gold Brick Hill. Its summit of about 1,950 feet, gave the enemy excellent observation of the land for miles. The Germans felt very confident that they could hold this position indefinitely. But supported with the fires of the 336th Field Artillery Battalion, the 3rd Battalion, 346th Infantry seized the hill.

One unusual aspect of this battle was the use of medium artillery to create craters on the hillside to provide shelter for the infantrymen to take cover in as they advanced. From the artillery's perspective, one of the highlights of the battle was the way in which communications and portable field radios, in particular, held up. Telephone wire remained intact despite heavy shelling and a large volume of armored traffic.[39] "The capture of the hill denied the enemy its observation . . . and broke the back of the enemy defenses in that sector of the Siegfried Line."[40]

By the first week in March, all three infantry regiments of the 87th Division had penetrated the last series of fortifications in the Westwall and were preparing to move east for their push to the Rhine. Before crossing that last natural barrier to the interior, the Golden Acorn Division captured one of the oldest cities in Germany. Situated at the confluence of the Rhine and Moselle Rivers, Koblenz had been the headquarters of the U.S. Army of Occupation following World War I and remained under American control until February 7, 1923, when France took over administration of the Rhineland. The 345th Infantry entered the city on March 17, 1945. Two days later, at 8:30 A.M., the ninety-five-man German garrison at Fort Konstantine surrendered, and after twenty-two years, the American flag once again flew over the ancient German city.[41]

At Koblenz the 87th Division artillery committed an act analogous to another nation destroying the Washington Monument, or at least similar to destroying Gen. Andrew Jackson's equestrian statue in the French Quarter in New Orleans. Artillery fire tipped over a gargantuan forty-five-foot statue of Kaiser Wilhelm I astride his horse located at Deutsches Ecke (Corner of Germany), that long, triangular strip of land in the city at the confluence of the two rivers. Years later an aerial observer, Phil Jackson of the 336th Field Artillery Battalion, claimed partial credit for the deed. Jackson insisted that it was not a voluntary act of vandalism but that he actually received orders from Brig. Gen. William Ford, who had received instructions from higher up the chain of command. The thinking was that destroying the statue would have a strong effect on enemy morale, which it probably did. Jackson remembered that a few direct hits from 105-mm shells only put a few dents in the statue. After a few more rounds from 155-mm shells struck the monument with little effect, he called corps and damage from their 8-inch howitzers effectively destroyed it.[42]

Over the next few days, the 87th Division pushed south toward Boppard to clear out the land between the Moselle and Rhine while the British and Americans gathered their units along the west bank of the Rhine to coordinate their efforts for the crossing. The 7th Armored Division had captured the famous Ludendorff Bridge at Remagen on March 7, allowing a number of First and Third Army units to cross the Rhine immediately. But in accordance with Eisenhower's broad-front strategy, the main American crossing took place around March 25.

At 10:30 P.M. on the twenty-fourth, a patrol from the 2nd Battalion, 345th Infantry crossed the river at Boppard to scout out Filsen, directly opposite the unit's designated crossing site. After crossing successfully while under enemy fire, the patrol came under heavy fire when they tried to return to Boppard. Because it was less than a half hour until the first assault wave was scheduled to cross, the men

The rebuilt equestrian statue of Kaiser Wilhelm I at the Deutsches Ecke in Koblenz. Photo by author.

The east bank of the Rhine River opposite Boppard, near the spot where the 345th Infantry made its crossing; note the steep incline. Photo by author.

simply stayed where they were. At the appointed hour, elements of the 2nd and 3rd Battalions of the 345th Infantry made the crossing unopposed, and by the early hours of March 25, Filsen was firmly in American hands.[43] Farther north the 1st and 3rd Battalions of the 347th Infantry began their assault crossing from the area near Rhens at the same time the 345th began theirs at Boppard. The 1st Battalion's objective was Ober Lahnstein, while the 3rd Battalion's goal was Braubach a little farther south.[44]

During the 347th's river crossing, the accompanying forward observers had to call for artillery fire from the boats in midstream. As both battalions of the regiment made their crossing, they were caught under illumination flares and received heavy fire.[45] "Every time a boat would make a dash for the opposite shore, the Jerry machine guns would follow his zig-zag course with tracers. When the smoke would begin to lift, the Jerry flak and artillery would again start pounding the men as they tried to lead in the boats."[46]

Despite heavy resistance, "all fighting battalions of the division were across the river by Monday night [March 26]."[47]

Prior to crossing the Rhine, the Americans might have measured their daily progress in terms of yards, but afterward they counted it in terms of miles. With the Russian army threatening Germany from the east and the Americans and British now across the Rhine, the Western Allies' advance into the interior of Germany greatly accelerated, with American units sometimes progressing as much as twenty to forty miles a day as German resistance rapidly deteriorated. "The Division advanced a total of over forty-five miles in the last seven days of March."[48] By the last day of the month, elements of the 345th Infantry were in the vicinity of Bodenrod, approximately sixty miles east of Boppard, as the pace quickened.[49]

The ability to provide spontaneous and effective close fire support required forward observers to move out front to find enemy targets. By positioning themselves ahead of maneuvering infantry, they incurred greater risks. Exactly where they placed themselves depended largely upon the features of the terrain. On April 10, 1st Lt. Ross H. Rasmussen was a forward observer from Battery A, 912th Field Artillery serving with a rifle company on the attack near Geschwende. After the riflemen became pinned down by intense machine-gun, small-arms, and *panzerfaust* fire, Rasmussen strapped on the heavy, bulky radio and moved out ahead of his infantry to a position offering better observation of enemy positions. From there he successfully adjusted artillery fire while exposed to intense German fire. As a result of his bold actions, four hostile machine guns were destroyed and a dangerous strongpoint reduced.[50]

The closing days of the war in Europe saw more examples of how the combat-arm designations of officers became blurred as artillery officers under fire assumed responsibility for infantry units. Also on April 10, 2nd Lt. Irwin R. Evens, forward observer from Battery B, 334th Field Artillery, was with an infantry company near Stutzhausen when the leading elements of the unit were stopped by an abatis. Immediately the company came under enemy automatic and small-arms fire. As the remainder of the men took cover, the lead platoon was cut off by two halftracks carrying more than two dozen SS troops. Evens crawled through some woods, and after finding ten of the men from the platoon, he attempted to return with them to the rest of the company. But they did not get very far before the

Germans spotted them and opened fire. After finding cover, Evens repeated his effort, locating an additional twelve men and returning all twenty-two back through the enemy's lines to their company, including two casualties. For his gallantry in action and superior leadership, Lieutenant Evens was awarded the Silver Star.[51]

Also performing double duty the next day was 1st Lt. John Connolly, a forward observer with Battery C, 336th Field Artillery. Connolly was accompanying Company I, 346th Infantry when enemy cannon fire stopped the men in their tracks. The lieutenant courageously moved forward and helped the platoon leader maneuver his men from the danger area without suffering heavy casualties. Later, when heavy enemy machine-gun fire delayed the company's advance, he courageously moved to a spot only fifty yards away from the enemy and calmly adjusted fire on the German position, destroying it.[52] Connolly, who would make the army his career, participated in so much intense combat during his overseas service with the 336th Field Artillery that his men admiringly gave him the nickname "Fire Mission John."[53]

April 1945 was a strange month for the soldiers of the 87th Division. Although it was apparent that the war in Europe was ending rapidly, no one knew how soon it would actually be over. After months of combat involving very slow progress, suddenly they were advancing rapidly. The Western Allies had found it necessary to form hard-hitting, highly mobile field commands able to reach into German rear areas, disrupt their communications, and disable their administrative units and installations.[54]

In most places German resistance was very light, but in a few others it was intense. By this time the Third Reich was using high-school boys and adult women to operate antiaircraft batteries. A communiqué addressed to the 87th Division Artillery in April indicates how depleted the ranks of manpower available to defend Germany had become, indicating that of seven prisoners taken from the 119th Panzer Grenadiers, three were boys under eighteen years of age with less than five weeks in the army. All of them were unarmed.[55]

Despite the apparent disintegration of the Wehrmacht, some German units were still willing to fight to the death. By April 7 the 2nd and 3rd Battalions of the 345th Infantry converged on Tambach, where a number of important highways intersected. In the predawn darkness on the morning of April 8, a group of Hitler *Jugend* attacked the 2nd Battalion in force. Company F bore the brunt of the assault, but when it was over, forty-four young Germans lay dead and many others wounded.[56] Years later Don Welever, who served with Battery B, 334th Field Artillery, observed that perhaps the most dangerous thing in Germany that month was a fourteen-year-old with a *panzerfaust.*[57] It certainly appeared to be true. Although the end was now only a month away, at Tambach "Company L lost eight killed and twenty-two wounded, including First Lieut. Hall, of Pensacola, Florida, its commander."[58]

The next day the 334th Field Artillery lost an entire forward-observer crew. Four men from Battery B became casualties as they were moving forward to relieve Lt. Gordon Howard, the forward observer with the 3rd Battalion. From machine-gun positions hidden along overhanging cliffs, the Germans ambushed Lt. Frederic Tower's party when the jeep driver, PFC Eddie Smith, took a wrong turn. Tech 4 Raphael Schoenberg and Pvt. Francis Fogle were instantly killed. Lieutenant Towers was hit in the chest but still managed to escape; he died two days later.[59] Smith was captured but survived.[60]

On April 17 the 347th Infantry captured what remained of the bombed-out city of Plauen, less than thirty miles from the Czech border; that afternoon the 345th Infantry moved in. Now it was clear that the war would end soon, but even so, no one could predict when. One thing was obvious, though: even if the Nazis continued to resist, Germany would soon be entirely under Allied control. Four days later the 334th Field Artillery Battalion was notified to expect contact with the Russians to its front.[61] This left the German army with little *Kampfensraum,* or space remaining to fight within the Reich.

Another indication that the end was near was noted in Capt. Henry Franey's last entry in the 334th Field Artillery's Battalion unit history for April: "This has been the least active month of the Bn's combat experience, in respect to firing. However, the rapid displacements necessary to maintain effective artillery support during the pursuit phase of operations following the Rhine crossing kept the Bn busy until April 17. The latter half of the month has seen practically no activity."[62] While the batteries were not completely inactive, the number of fire missions tapered off significantly. Between April 18 and April 30, the 334th Field Artillery fired one short

German civilians walk among the ruins of Plauen, May 1945. The sign reads "For Pedestrians." Photo courtesy of Donald L. Walker.

German soldiers near Plauen after the armistice, May 1945. Photo courtesy of Donald L. Walker.

time-on-target mission on April 19, another the next day, harassing and interdictory fire on the night of April 27–28, and another short time-on-target (on Arnoldsgrun) on the last day of the month.[63] During the first week of May, the number of fire missions decreased even further. Throughout this period the battalion expended only fifty-three rounds.[64] Even at this time, there was no clear indication that the war in Europe would be over by the end of the week. On May 5 a message was sent to the 334th Field Artillery regarding an impending limited-objective attack to take place the next day. The battalion moved three times, first to the vicinity of Arnoldsgrun, then near Schoneck, and finally close to the tiny pastoral village of Kottenheide. From the latter it had gun positions that could reach targets across the Czech border. Even on the morning of May 7, the battalion dispatched three forward-observer parties to accompany

87th Division 105-mm howitzers near Kottenheide after the war ended. From here the guns could reach targets in Czechoslovakia. Photo courtesy of Donald L. Walker.

the 3rd Battalion, 345th Infantry, but at 5:45 that morning received this message: "Cease firing on all missions by order of the Division Commander."[65] The war in Europe was over.

In contrast to the German army's last-ditch counteroffensive in the Ardennes, the final months of World War II saw the 87th Division attack into Germany itself. In early 1945 the Golden Acorn Division had to confront a tenacious and still very effective enemy struggling to repel an invasion of its homeland. A large part of the German army's remaining effectiveness was its capacity to wage combined-arms warfare as it fell back, first to the border, then ever deeper into the Fatherland. As demonstrated for more than a year in Italy, German effectiveness in defense was due in large part to their capacity to make the best use of terrain, especially to take advantage of high ground, and in their command over, control of, and communications with their own defending troops. But by mid-February major

Lt. Don Walker by a Stuka, most likely at the Plauen Airfield, May 1945. Photo courtesy of Donald L. Walker.

problems had developed with the German army's ability to maintain effective communications, particularly in the American Third Army's sector.

As the Allies approached victory, it was clear that the 87th Division had achieved a more effective coordination of infantry and artillery than ever before. As many incidents during these months reveal, forward-observer personnel of all ranks went beyond the performance of their artillery functions as circumstances dictated. In the process, they became fully integrated as fighting men on the front, exposed to the same dangers and sharing the same risks as the combat infantrymen they were supporting.

The artillerymen of the 87th Division saw a repeat of two critical lessons. First, forward observers often had to drop their designation as artillerymen and transcend their training to become leaders at the front line of combat. In crisis situations they accepted responsibility for the men around them and the completion of their "combined" mission, not just their task as observers of enemy targets and directors of artillery fires. Of course, self-preservation played a part in this. The second lesson relates to the enormous cost in manpower brought by combined-arms warfare. As officer observers became fewer and fewer, observation teams came under the direction of NCOs and lower- ranking enlisted men, who shouldered the responsibility for observing and detecting the enemy and adjusting fire without causing friendly casualties. In the European theater, just as it had in the Pacific, waging effective infantry-artillery warfare meant not only a blurring of responsibilities between combat arms but also a blurring of the ranks of artillery.

| # Conspicuous Gallantry

Forward Observers and the Medal of Honor

Medals and ribbons are an important way of raising morale and creating esprit de corps among members of the military. More than that, the citations for medals represent a primary source of information for military historians by recording a significant event that took place at the individual level during a war. While medals are awarded on an individual basis, the action the recipient takes generally affects the lives and welfare of many of his comrades. While many if not most individual acts of heroism may not turn the tide of battle, each makes a contribution in its own way. Many who survive a war may owe their lives to one who did not. Thus, the citation for a medal represents more than just a tribute to the recipient—it documents a smaller facet of a larger historical event.

In the absence of diaries, personal accounts, and interviews, perhaps the only practical alternative to gaining a clear understanding of what forward-observer personnel contributed during World War II is to look at the citations for the medals they received. The after-action reports, unit histories, and other official U.S. Army records rarely make any specific mention about the actions of forward observers, and those that do rarely cite individuals by name. Because the general orders of the 37th Division in the Pacific and the 87th Division in Europe indicate that forward observers in those units were frequent recipients of medals, this is a reliable indicator that such men in all divisions of the army followed basically the same pattern of action during World War II.

Prior to 1970 the army's four highest awards for valor were the Medal of Honor, followed in order by the Distinguished Service Cross, the Silver Star, and the Bronze Star, which may be awarded for either heroism or service. When the army awards a Bronze Star for heroic action, it typically writes a narrative citation to describe the circumstances of the deed, which is then included in the military orders that create the award. As a rule, when the Bronze Star is awarded for meritorious service, there is no narrative. But some of the 37th Division's general orders for Bronze Stars include citations for meritorious awards. For example, General Order No. 46 awards the Bronze Star to Sgt. Robert S. Harshbarger of Alliance, Ohio, and reads: "for performance of *meritorious service in action* against the enemy at New Georgia, Solomon Islands, on 5 August 1943," while leading an artillery forward-observation party. But based on what Harshbarger did to earn his medal, his actions typically would be described as heroic on most citations: "When a strong patrol of the enemy attempted to ambush his party, Sgt. Harshbarger coolly opened fire with his sub-machine gun and kept the enemy pinned down until his comrades had safely withdrawn, killing two and wounding several of the enemy."[1]

In 1947 the army retroactively awarded the Bronze Star to all soldiers who had served in World War II who had received either the Combat Infantryman Badge or the Combat Medical Badge. For these retroactive awards, there are no individual citations because they were made across the board to all who met the criteria regardless of what each man had actually done. Many veterans took offense to this action. One combat veteran wrote to the editor of *Infantry Journal* complaining because he felt that too many Bronze Stars for meritorious service had already been awarded to rear-echelon troops during the war, while others had, as historian Gerald Astor has suggested, gone to the front, spent the minimum length of time required in a quiet area, and returned to the rear, where they then completed the papers necessary to receive the Combat Infantryman Badge. Harold Gordon complained to the editor of *Infantry Journal* about the decision to award the Bronze Star to all former combat infantrymen, writing: "This was and will be unfair to the officer or EM who had actually earned his medal or award the hard way. If you want the Bronze Star to be the equivalent of the Good Conduct Medal, all right, but then don't pretend that it means anything."[2]

All this may sound like much ado about nothing to those who have never served, but a military uniform is almost like a visual-display resume that the individual wears, indicating pertinent aspects of the serviceman's military career. Medals, badges, and ribbons do mean something to veterans and perhaps even more so to career military personnel. More than that, among veterans, particularly those who served in time of war, there is unspoken but conscious deference and respect to those who have been involved in combat. In general, 75–80 percent of all troops in the U.S. Army since World War II have been service personnel. Certainly, cooks and company clerks became riflemen during the Battle of the Bulge, but that is not typically the case. The public perception of veterans is that they all defend our country. While every soldier's service is important and most veterans' service is honorable, only a minority of all soldiers take part in any combat, an even a smaller percentage routinely. Accordingly, medals do mean something to military personnel and veterans.

The higher the award, the more deserving it appears to be, but there are always exceptions. An early postwar edition of *Infantry Journal* reported, "In the Southwest Pacific early in the war, the Distinguished Service Cross, specifically created to reward 'extraordinary heroism,' was passed out to some generals who had not been within fifty miles of the enemy."[3] The same article explained: "With few exceptions, the Medal of Honor was preserved as the one decoration that couldn't be kicked around even by the commanders of highest rank. Perhaps the reason is that this medal is the only decoration which cannot be awarded by any headquarters other than the War Department."[4]

The Medal of Honor is the highest military decoration for bravery that can be awarded to any member of the U.S. armed forces. It is presented by the president "in the name of Congress." As Gen. Bernard Trainor observed: "A junior enlisted man wearing . . . [one] is held in awe even by generals and admirals. They conjure up images of what it must have taken to win that coveted award and, in the process, wish they had one."[5] General Patton once remarked, "I would give my immortal soul for that award and President Harry Truman often told the men to whom he present[ed] the medal, 'I would rather have that medal than be president of the United

States.'"[6] Congress awards the medal for an act of personal bravery or self-sacrifice above and beyond the call of duty by a member of the armed forces in action against an enemy of the United States. Yet originally, it was not awarded according to this criteria.

First given out in 1863, the army awarded 1,200 of these medals during the Civil War and an additional 524 in the intervening years prior to World War I. It then awarded 95 during World War I and an additional 293 during World War II, a combined average of about six per month. During the Korean War, only 78 soldiers received the Medal of Honor, and 154 did so during Vietnam. So during these two limited conflicts, the army awarded just two medals per month on average. Thus, the total number awarded for the four major conflicts of the twentieth century was 610, or a little more than half of the number awarded during the Civil War.[7] This was because Congress awarded the first medals for such things as an extension of service or the saving of lives under noncombat conditions. After two years of reviewing the criteria for the award, "on July 9, 1918, Congress set standards that would ensure the Medal of Honor would only be awarded in truly exceptional instances to members of the Armed Forces who . . . in action involving actual conflict with an enemy, [distinguishes] himself conspicuously by gallantry and intrepidity at the risk of his life above and beyond the call of duty."[8]

In their respective lifetimes, if not still today, the names of Medal of Honor recipients Sgt. Alvin York and Audie Murphy were household names. In contrast, except for their families and perhaps some former comrades, few people are familiar with the names of Forrest Peden; James Robinson, Jr.; Lee Hartell; and Harold Durham, Jr., or that each earned the Medal of Honor. The purpose of recalling here what these four artillerymen did to earn the nation's highest award for valor is not to celebrate their deeds as forward observers but to commemorate their personal sacrifice. The army made all four awards posthumously.

Forrest Peden

Tech 5 Forrest E. Peden, a native of St. Joseph, Missouri, was a field artillery forward observer serving with Battery C, 10th Field

Artillery Battalion, 3rd Infantry Division. His division probably served in more theaters of action and was in combat for a longer duration than any other command in the U.S. Army during World War II. Landing near Casablanca in November 1942, the 3rd Division fought in North Africa, Sicily, and Italy before landing on the eastern coast of France in August 1944 and fighting its way through Lorraine and into Germany.

By early February 1945, elements of the 3rd Division were about sixty miles from the departmental capital of Lorraine, Strasbourg, and close to the Rhine River. With intentions of encircling the town of Neuf Brisach and sealing off the western bank of the Rhine in that sector, the 2nd and 3rd Battalions of the 7th Infantry made an attack from the north toward Biesheim at 2:30 on the morning of February 3. But the Germans were waiting for them. Outnumbering the Americans four-to-one, they sprung an ambush, unleashing a fury of artillery, mortar, machine-gun, and small-arms fire on the advancing infantry. The Americans took cover in a ditch, which they found to be occupied by the enemy.[9]

The first thing Peden did beyond his forward-observation duties was to go to the aid of two wounded soldiers and administer first aid under heavy fire. Realizing that his unit's radio was inoperative and that without it the Germans would quickly overrun them, Peden raced back 800 yards under intense fire to reach the battalion command post. By the time he got there, his jacket had a number of bullet holes. Next he found two tanks to join the battle in progress. To expedite their arrival, he climbed on the lead vehicle and directed it back to the scene. As the reinforcements drew closer, they attracted an increasing barrage of German artillery and small-arms fire. Just as the lead tank was ready to fire, a direct hit turned it into a blazing inferno, instantly killing the young artillerist. Aided by the light of the burning tank, additional reinforcements arrived and were be able to drive off the enemy.[10] Peden's actions were responsible for thwarting a German ambush, enabling the 7th Infantry to successfully carry out its mission to capture Biesheim—a small thing, perhaps, in the bigger picture of the war, but more importantly, his actions saved the lives of many of his comrades.

James E. Robinson, Jr.

Also serving in the European theater, 1st Lt. James E. Robinson, Jr., of Toledo, Ohio, was a forward observer with Battery A, 861st Field Artillery Battalion, 63rd Infantry Division. The 63rd like the 87th Division did not get into the war until December 1944, landing at Marseille on the eighth. From there the troops were transported north by train. By the end of the month, two regiments of the "Blood and Fire" Division were in line north and east of Strasbourg, not far from the same area the 87th Division had just departed days earlier, while the 254th Infantry was supporting the 3rd Division near Colmar.[11]

During the month of January 1945, the division remained in that part of Lorraine bordering Germany near Saarbrücken. At one point another regiment even used the same location for a command post that the 87th Division's 345th Infantry had occupied at Moranville Farm eighteen days earlier. By the beginning of March, the 63rd Division was crossing the German border east of Saarbrücken, and by the Ides of March had breached the Siegfried Line east of Fechingen by Ensheim. On March 28 the division crossed the Rhine in the general vicinity north of Worms by Hamm. By the end of the month, the "Hotshots" had passed by Heidelberg, crossed the Neckar River, and were moving on line eastward north of Heilbronn.

On the morning of April 4, the 1st and 2nd Battalions of the 253rd Infantry crossed the Jagst River under a heavy barrage of German artillery and mortar fire. The 2nd Battalion immediately moved out to the southwest, while the three rifle companies of the 1st Battalion spread out to the east along the south bank of the river. Company A moved into the town of Untergriesheim.[12]

By April 6 Company A had moved across some open fields and taken the high ground before the village of Kressbach. Here the company suffered many casualties and lost all of its infantry officers to ongoing enemy artillery and mortar fire as it pressed to the east. Lieutenant Robinson, the forward observer attached to Company A, lost both of his crew members to the German fire. Battery A sent two enlisted men to replace them, but both were hit by small-arms fire and evacuated, leaving the lieutenant to carry the seventy-eight-pound radio and also leaving him in charge of the infantry company.[13]

Going beyond his duties as forward observer, Robinson gathered the nearly two dozen uninjured men remaining plus a few walking wounded and rallied them to resume the attack. While carrying the heavy field radio, he led the American advance, killing ten German soldiers in foxholes with rifle and pistol fire. His men then swept the immediate area of further resistance. Soon afterward the lieutenant was ordered to assault the town of Kressbach. During the advance that followed, a shell fragment struck him in the throat. Although he was in great pain and had lost a great deal of blood, he ignored medical attention and pressed on with the attack, adjusting supporting artillery fire despite his severe wounds. When the town had been taken and he was unable to speak, Robinson left the group he had inspired and walked nearly two miles to reach an aid station, where he died from his wounds.[14]

Years later, SSgt. Paul W. Vermillion of Port Lavaca, Texas, a veteran of Company A, remembered that he had been very close to Lieutenant Robinson at the time he was wounded, perhaps ten to twelve feet away, and recalled: "It was like someone had poured about a gallon of blood down his neck and chest. He was walking toward Kressbach. I have it from a reliable source—the reason he died at the aid station was that they had a tube in his throat to clear the blood drainage & Lt. Robinson pulled the tubing out & the result was that he probably drowned from his own blood."[15]

None of the forward-observer personnel with Company A came away unscathed that day. Vermillion emphasized this, adding: "Lt. Robinson had two men with him from A Battery, 861st Field Artillery. One was T/5 Charles T. Moody who was K.I.A. The other was PFC Tackett who lost his testicles and penis to a wound. The A Battery, 861st Field Artillery Commander—Captain (later B/General) Charles 'Pop' Young—said he sent two men as replacements & they were also wounded."[16] While the full extent of Robinson's contribution may be hard to determine, the history of the 63rd Division records that it enabled elements of the 1st Battalion to enter and successfully hold Kressbach.[17]

Lee R. Hartell

Only one field artillery forward observer earned the Medal of Honor during the Korean War, and he was also a veteran of World War II. Among the enlisted field artillerymen with the 43rd Division, the Connecticut–Rhode Island National Guard unit, in the South Pacific was Lee Ross Hartell. Born in Philadelphia, Hartell was a member of a battery's survey section who served on both New Georgia and the Philippines. After the war Hartell reenlisted in the guard and, after passing tests to receive a commission, became a second lieutenant. Shortly before Christmas 1950, he became a member of the 15th Field Artillery Battalion, 2nd Infantry Division then fighting in Korea.[18]

Some historians have described the fighting during the Korean War as being similar in some respects to the immobile warfare of World War I. That is, after an initial period of relative mobility, the fighting settled down to combat between two stationary front lines. Referring to the situation during the winter of 1950–51, the history of the 2nd Division indicates that "the static situation on the central front had permitted construction of a well-fortified MLR [main line of resistance]" and that "from an overall viewpoint, the situation remained quiet throughout the winter and early spring. To the men who patrolled across the snow-covered rice paddies, however, the war seemed far from quiet."[19] During the first half of 1951, Lieut. Hartell flew more than 200 aerial observation missions.[20]

By July 1951 the 2nd Division, still north of the thirty-eighth parallel, had moved to a new sector, with the Yokkok River on its right and overlooking the corridor that led to the city of Chorwon. It was a period of relative inactivity except for aggressive patrolling and the reconstruction of many defensive works that the rainy season had destroyed.[21] Over the summer, Lieutenant Hartell had the chance to go home but declined. After another forward observer at the front was killed, Hartell volunteered to go as his replacement, arriving on the night of August 26.

That same evening Company B, 9th Infantry Regiment received orders to occupy Hill 700 in front of the main line of resistance, which it did. Early the next morning, the Chinese counterattacked the hill from all four sides of the company's position.[22]

As the infantry unit's accompanying forward observer, Lieutenant Hartell quickly called for flares to illuminate the enemy's avenues of approach, then directed very effective fire on them. Soon afterward a large number of enemy troops swept up the hill to a point within ten yards of Hartell's position. Despite suffering a severe wound to the hand, the lieutenant used his other hand to hold the microphone and continued to adjust fire with great precision and effect. As a result the enemy temporarily fell back, then returned in even greater numbers. With a numerically superior force sweeping over the outpost and about to overrun his position, the lieutenant made a final radio call urging both batteries to fire continuously. Despite the severity of his wounds, Hartell's courageous actions contributed greatly to stemming the enemy tide and enabled his company to hold on to its strategic strongpoint.[23]

Harold B. Durham, Jr.

Although many U.S. Army veterans of World War II took part in the Korean conflict, a surprising number of them also served in the Vietnam War. Harold B. Durham, Jr., who would become the only field artillery forward observer to receive the Medal of Honor for service in Vietnam, was not yet four years old when World War II ended. He acquired the nickname "Pinky" at birth when the hospital in Rocky Mount, North Carolina, where he was born ran out of blue blankets for newborn baby boys.[24] Durham enlisted in the army in 1964 and served in Vietnam as a rotary-wing maintenance mechanic. During this first tour of duty, he also volunteered to teach English in a nearby village. Offered the chance to attend Officer Candidate School, he accepted and graduated on December 17, 1966. After receiving his commission as a second lieutenant of artillery, he voluntarily returned to Vietnam. During his second tour of duty, Durham was assigned to Battery C, 6th Battalion, 15th Field Artillery, 1st Infantry Division.[25]

The "Big Red One" was one of the first divisions deployed to South Vietnam in the summer of 1965 after Gen. William C. Westmoreland received authorization to begin a massive buildup of army troops there. With its headquarters base camp at Di An, only eighteen miles northeast of Tan Son Nhut Air Base and the South Vietnamese capital of Saigon, the 1st Division operated in War Zone D, conducting a number of search-and-destroy operations and sweeps during the time that it was there. Between September 27 and November 19, 1967, the 1st Division conducted Operation Shenandoah I and II in Binh Long and Binh Duong Provinces to find and destroy Vietnamese Communist strongholds.

On October 17, elements of the 721st Vietcong Regiment ambushed and trapped the 1st Division's 2nd Battalion, 28th Infantry at Ong Thanh, about twelve miles northwest of the division base camp at Lai Khe, approximately forty-one miles north of Saigon. Artillery and airstrikes ultimately enabled the Americans to fall back, saving them from total annihilation, but losses were heavy, including seventy-two wounded, fifty-two killed, and two missing. The U.S. Army never got an accurate count of the total enemy insurgents killed, and the true number may never be known. Nevertheless, it recorded 141 confirmed and 121 unconfirmed enemy killed during the battle.[26]

Prior to October 17, a different team, a lieutenant and his sergeant, had served Company D as its forward-observer party. But shortly before the date of the ambush, during a lull in activity, the two artillerymen sat side by side cleaning their .45-caliber automatic pistols. When the sergeant was finished, the lieutenant asked to see the gun and yanked it out of his hand. The weapon discharged, fatally wounding the NCO in the stomach.[27]

Company D's commander, Lt. Clark Welch, had witnessed the careless act and ordered the artillery lieutenant to leave on the same helicopter that carried the sergeant's body. The next chopper in brought with it a new forward observer, Lt. Harold "Pinky" Durham, who had volunteered for the job.[28]

About midmorning on October 17, the Black Lions made contact with a Vietcong force both in the trees and in well-concealed, fortified bunkers. As the battle began, Lieutenant Durham quickly moved to an exposed position and began adjusting supporting artillery fire. At one point during a lull in the fighting, Pinky administered first aid to the wounded despite being subjected to heavy enemy sniper fire.[29]

As the intensity of the combat increased, Lieutenant Durham took a position near the battalion commander, Lt. Col. Terry Allen, Jr. At this time Jim Gilliam, Durham's radiotelephone operator, witnessed the explosion of a rocket grenade that knocked the lieutenant off his feet, leaving his glasses dangling from his left ear. When he tried to put them on straight, they dangled from his right ear, suggesting that one or even both ears may have been partially blown away. "Someone on the other end of the radio wanted to stop the artillery fire to bring in more air, but Durham, with Lieut. Welch at his side, vehemently disagreed. 'No,' he said, 'Hold that air strike. Keep sending us artillery. It's what I need. I know what I need.'"[30]

Soon after this, Durham learned that Company A, while bearing the brunt of the ambush, had lost its forward observer.[31] The wounded artilleryman, a Lieutenant Kay, was in bad shape. With blood all over his face, his leg smashed, a big chunk of flesh missing from his wrist, and two gashes in his shoulder, the officer was unable to call for fire missions.[32]

Thinking that he could be more effective with Company A, Lieutenant Durham instructed Gilliam to remain near the battalion command post while he took off for the Company A front. Once there, he moved the artillery fire to the eastern edge of the battlefield, trying to seal off the Black Lions from the enemy to give them time to remove their wounded.[33] As Durham was running to take the place of Kay, the Vietcong detonated a claymore mine, severely wounding him in the head. Despite his intense pain, he kept adjusting artillery fire while firing his own weapon to support the overwhelmed infantrymen. As the enemy closed in, Lieutenant Durham directed the supporting fire to be placed almost on top of his own position. This worked, and twice the attackers were driven back, leaving many of their own casualties behind.[34]

By this time Durham was in extremely weak condition, yet he continued to adjust fire where it was needed most. Refusing to find cover, he instead found a small clearing that gave him a better view for adjusting fire. But he was badly wounded again, this time by enemy machine-gun fire.[35]

The company commander, Lieutenant Welch, who had suffered multiple wounds, including most of one bicep shot away, and was weak from loss of blood, realized that by now, as he made this transmission, Durham was operating his radio 'press to talk switch' using a bloody stump where his hand had been.[36] Lieutenant Durham clung to life long enough to perform one more heroic act before succumbing to his wounds. The concussion from a rocket-propelled grenade hitting a nearby tree momentarily knocked unconscious Company D's first sergeant, Bud Barrow. As he regained consciousness, Barrow saw Durham waving to him as he was trying to rise. "Top," the lieutenant called out, pointing to two enemy soldiers approaching with AK-47 assault rifles. As Durham fell for the last time, the sergeant spun around and fired, his last word had saved Barrow's life.[37]

Lieutenant Welch had been sitting against a tree and could actually see artillery shells coming through the tops of nearby trees and the flash of explosions all around. After gathering his strength, he dragged himself through the grass around the perimeter but found few surviving Black Lions.[38] Even for a rifle company at full strength, 155 casualties would be a tremendous loss. Artillery and airstrikes played a role in driving back the enemy and saving American lives. How many lives Harold Durham saved by his actions other than Sergeant Barrow's is indeterminable. But the motto of the 1st Infantry Division is "no mission too difficult, no sacrifice too great; duty first." Lieutenant Durham certainly lived up to that motto, making the ultimate sacrifice, as had Tech-5 Forrest Peden, Lt. Jim Robinson, and Lt. Lee Hartell.

Conclusion

The creation of the FDC, the firing chart, generally reliable field radios, and the newly improved ability to mass fires developed in the years between the two world wars certainly enabled U.S. field artillery to realize the full potential of its combined-arms capability in these three wars of high-intensity combat. The number of artillery strikes directed by aerial observation far outweighed those done from the ground after World War II. Yet without trained artillerymen accompanying maneuvering infantry on the ground, the kind of interaction that took place between these four Medal of Honor recipients and the men they were supporting, along with the contributions each made, could not have taken place. These four Americans demonstrate, perhaps to the extreme, the extent of a forward observer's willingness to transcend his duties as an artilleryman and become deeply involved in the combat going on around them.

CHAPTER 11

The Key to Successful
Infantry-Artillery
Coordination

The evolution of the forward observer and the establishment of forward-observer teams at the outbreak of World War II provided the human element necessary in the development of an effective combined-arms tandem of field artillery and infantry; it was the vital link between the two. Other combinations of combat arms developed varying levels of cooperation and coordination as the war progressed, but as John J. McGrath has noted: "U.S. field artillery in World War II was an unmitigated success in all theaters from the beginning to the end of the war."[1] Russell Weigley underscored this point when he commented, "the artillery was the American army's special strong suit," and General Patton even suggested that field artillery had won the war.[2]

Combined-Arms Execution: The Big Picture

Regarding tactical air support, the rigid command structure in place in the early stages of the war apparently hamstrung effective results. "Army units found that air support took so long to arrive that there was little point in requesting it."[3] American air power in the Pacific in 1943 did a wonderful job of turning Japanese rear areas into a "shambles" but did little to give friendly ground forces close support. Even when they tried, they sometimes hit Allied troops without inflicting much damage on the Japanese, making close air support a "negligible factor" in the battle for New Georgia. Air-ground coordination on Bougainville was thought to be somewhat better and proved that bombs could be dropped closer to friendly troops than previously believed. But even the most accurate bombing could not destroy many dug-in Japanese infantry positions. This and the large number of calls for aircraft resulted in a long delay between the request for and the air arm's delivery of a strike. Yet by the time of the Luzon invasion, when a sufficient supply of tactical aircraft was available, the gap in time between the request for help and its delivery was often reduced to a few minutes.[4]

Regarding the 1943 invasion of Sicily, Gen. Omar N. Bradley commented: "In vain, we searched the skies for close air support from our airmen."[5] During the Normandy campaign, the weather and the difficulty in identifying targets on the ground hampered close air support. Target identification was easier when the enemy was moving in daylight. Some German officers considered Allied artillery fire to be more dangerous than Allied aircraft.[6]

But American air-ground cooperation matured during the Normandy campaign, mainly out of necessity. Shortages of artillery ammunition did away with the previously held theorem that targets in range of the guns should be left to artillery. The results were surprisingly good. Air-ground cooperation improved as the two combat arms became better acquainted with each other.[7]

Close air support had the capability of wreaking widespread devastation on enemy personnel, in particular infantry. This tremendous lethal power was a two-edged sword, however, as demonstrated during Operation COBRA, during which the United States suffered over seven hundred casualties from friendly fire. Although airplanes and weaponry would change, "the basic approach to close air support developed in the ETO [European theater of operations] during World War II has continued to the present day."[8]

The development of American infantry-armor coordination seems to have come at about the same pace as that of air support. Tactical mobility improved as infantrymen were able to assist tanks in eliminating increasingly sophisticated antitank devices while tanks were able to protect infantrymen from enemy armor.[9] Just like its air support, infantry had to become better acquainted with its armor support. In June many of the U.S. infantry divisions in Normandy were new to combat and had never trained extensively with their supporting tank battalions. Certainly opinions may vary, but just as it had with the Army Air Corps, infantry developed a much closer working relationship with its armor support during the breakout from Normandy, particularly when a tank battalion remained in support of the same infantry division.

Yet throughout the entire war, the radios issued to infantry, armor, and tactical aircraft had different frequencies, making communications between the arms impossible.[10] Don Welever was a member of a forward-observer party with a radio. But to be able to communicate with one American tank crew in Germany in February 1945, Welever had to pound on the hull of their turret to prevent them from firing on friendly troops.

Disparity in the Pacific, Parity in Europe

At the tactical level, the experiences of the 37th and 87th Divisions were, on balance, not identical. Although there were many similarities in the problems of command, control, and communications, the two divisions were fighting enemies who were quite different in their tactical approach to warfare. Yet both divisions were able to execute a high degree of coordination between infantry and artillery and mass supporting fires with great effect.[11] Both divisions carried out their primary artillery mission very effectively.[12]

In the Pacific the 37th Division was fighting a much-less-sophisticated enemy. Japanese doctrine emphasized manpower and aggressiveness on the offensive, a perspective rooted in that nation's faith in the superiority of its own infantrymen. Thus, Japanese infantry doctrine was stuck in outdated nineteenth-century thinking and never developed a combined-arms practice to the degree that the Western industrial powers did during the interwar era (1919–39). Yet whether one calls it fanaticism or an unquestioned obedience to duty, the highly aggressive style of Japanese fighting created its own unique set of difficulties for the Americans in the Pacific, while lending itself to distinct tactical advantages for the United States.

In contrast, Germany had developed its tactical doctrine to a very sophisticated level by the time the United States entered World War II. German faith and expertise in technology and their developments in *bewegungskrieg* (mobile, or open, warfare) allowed them to build a formidable war machine; their understanding of and experience on the battlefield (a product of an earlier war and two decades of interwar study) led them to create their own version of combined-arms doctrine. Thus, the Wehrmacht, with generally more combat experience than its American counterpart, did not typically pit its tactical weaknesses against American strengths. In addition, most German soldiers had no great desire to kill as many Americans as they could before dying.

Soldiers of the 87th Division routinely confronted infantry and armor assisted by supporting artillery directed by skilled German forward-observation teams. The division's struggle demonstrated the general parity existing at the level of tactical doctrine in the war in Europe. Until the very last months of fighting, the German army was able to apply firepower to support its infantry and armor in retreat, defense, and counterattack, complicating the 87th's mission.

The Vital Role of the Forward Observer

British military historian Jonathan B. A. Bailey sees World War I as "the birth of modern warfare." With it came a shift in military doctrine ushering in the standard use of indirect fire. Bailey describes that shift as "the key innovation [to] a new approach to combat founded on indirect fire."[13]

The use of indirect artillery fire requires an observer to conduct and adjust fire. The forward observer shouldered responsibility for control over the application of firepower on the battlefield. The actions of such men with both the 37th and 87th Divisions demonstrate conclusively a remarkable improvement in the ability of artillery to provide close support to maneuvering infantry at the most critical times compared to what it had been able to do during World War I.

After 1918, tactical doctrine for field artillery changed to the extent that the forward observer and the fire-direction center split the duties formerly carried out by a battery commander. The ability to provide close support worked during World War I as long as the locus of action remained immobile. Once targets became removed from the sight of forward observers and battery commanders, their ability to provide observed fire ceased. But once observers began to accompany maneuvering infantry, it enabled field artillery to make what has been described as "a quantum leap forward in its ability to participate in mobile warfare."[14]

During World War II, the forward-observer team had two distinct tasks: first, to detect the enemy and identify targets, and second, to control the artillery fires, whether (as in most cases) by adjusting indirect fire over great distances or, in the case of the 37th Division in Manila, by controlling direct fire on targets in front of them, and to adjust those fires to achieve maximum effectiveness. Terrain, vegetation, climate, and other factors on the specific battlefield, such as the nature of the target, shaped a team's ability to accomplish its mission. These factors came into play whether ground- or aerial-observation methods were best suited for the targets at hand.

The experience of the 37th Division shows that aerial observation was best for targets in the open or moving targets; ground observation was essential for detecting and adjusting fire on an enemy in concealed positions.[15] The terrain and vegetation of the Pacific islands forced the forward observers to become adept at controlling artillery firepower by sound when control by sight proved impossible.[16]

Aerial observation used for target acquisition and artillery adjustment during World War II ordinarily had many advantages over ground observation due to the speed and mobility of an observer in an airplane as well as the more expansive view afforded from altitude. Barring poor visibility and flying conditions, these advantages in finding targets and adjusting fire typically outweighed those derived from ground observation. Both airborne observers and forward observers on the ground performed a dangerous duty. As agents of destruction, both played a vital role in providing effective artillery support. But directing artillery strikes from the air had one limitation that ground sensing did not—weather. Although the conditions necessary to conduct effective strategic bombing and to adjust artillery fire from the air are not identical, a study completed after the war noted that weather conditions reduced the utility of strategic-bombing operations in Europe by as much as 25 percent. The study also found that during the winter months, as many as ten to fifteen days a month were nonoperational.[17] Bad weather prior to and during the initial phases of the Ardennes offensive limited aerial reconnaissance as well as artillery adjustment, increasing the importance of having forward-observation teams on the ground in that bloody and extended campaign.

The Evolving Responsibilities of the Forward-Observer Team

The records of both divisions demonstrate that combining combat arms on the modern battlefield necessitates the integration of artillerymen with those at the front lines on the battlefield, whether for indirect or direct application of artillery fire. In that integration on the battle line, designations of specific combat arms become temporarily meaningless. To practice combined arms effectively required modern armies to blur the distinctions between infantry commanders and artillery observers, and prepare the latter to be officers both in command of troops and in control of firepower simultaneously as necessary. This is exactly what the Officer Candidate School for Field Artillery at Fort Sill emphasized during World War II—leadership skills. The numerous examples of these actions by artillery forward observers, acting in crisis situations and shouldering responsibility for leading infantrymen at the front, provides evidence for this necessary integration of "command" and "control."

Not only a blurring of the distinction between combat arms but also the distinction among the ranks broke down as World War II slowly drew to a close. The lack of commissioned officers forced NCOs and other enlisted men to assume responsibility for leading forward-observation teams and even leading infantry troops on the increasingly integrated battlefield. The experience of the 87th Division in the costly struggle to breach the Siegfried Line and cross the Rhine River demonstrated the degree to which young enlisted men took the initiative and bore the responsibilities usually assumed by commissioned artillery officers.

Communications: The Process linking Control to Command

In both the Pacific and European theaters, climate and terrain were the principal or determining factors shaping communications between forward observers and their artillery battalions. In the Pacific, moisture and humidity severely hindered reliance on radios, telephones offering the most reliable means of communication between observers and artillery battalions. But as trails were cleared to transport supplies forward and later widened to make roads, construction crews frequently severed the observers' telephone lines.

Despite problems with radios in Europe, the climate there was much more conducive to their use. In this more evenly matched war, however, the 87th Division had to contend with different aspects of German radio technology—the ability of the enemy to monitor U.S. radio traffic and not only to listen in but also to determine the approximate location of the origin of broadcasts. Correspondingly, Allied technology could intercept German transmissions and determine enemy intentions at key moments during the war's last years. An additional hazard associated with the forward observer's use of radios was the antenna. It was particularly conspicuous and certain to give away the location of the radio operator, and everyone with him, if not carefully hidden.

How Forward Observers Were Perceived by Friend and Foe

Like all other soldiers performing their jobs or routinely serving under fire during World War II, not all forward observers were particularly brave, did their jobs particularly well, or risked their lives more than necessary. They were, however, part of that very small group out of the entire army who routinely engaged in combat. They lived, bled, and died with the infantrymen. Some have argued that because forward-observer personnel were relieved more frequently than the combat infantrymen, their experience did not compare. Perhaps that was because some were able to return to their batteries for rest and hot food more frequently than the riflemen could get relief from the front lines. But forward observers in World War II were in short supply, limiting their opportunities to return to their batteries during combat. Capt. Hugo Gisske of the 912th Field Artillery observed that even when an infantry regiment relieved another, the forward observer usually had to stay on with the new unit. This was because he already knew the front and his battalion's batteries were already registered on specific points along it. Bob Booth's experience at Bonnerue, Belgium, serves to corroborate this for servicemen in Europe.[18] Forward-observation duty also entailed an additional liability, for these men were high on the enemy's priority list of human targets. The long radio antenna and binoculars distinguished them as important targets.

Riflemen did not universally hold forward observers in high regard. Bad ones were particularly loathed. Soldiers on both sides feared the destructive power of skilled enemy observers and in some instances even hated them. Ed Laughlin, who served with the 82nd Airborne in the Battle of the Bulge, recalled seeing the suspended corpse of one German forward observer the Americans had executed by hanging him with his belt after he tried to surrender. He had been directing artillery and mortar fire on the soldiers with deadly accuracy, killing and maiming their comrades.[19]

Most forward observers were ordinary men—soldiers doing their assigned jobs as best they could. Infantrymen of the 96th Division serving in the Pacific had high praise for their forward-observer personnel, asserting that they "received shamefully little recognition for their heroic and dangerous deeds in combat. . . . These same sections worked shoulder to shoulder with front line companies, many times with the farthest advanced elements of the infantry. They suffered everything endured by the doughboys and with them brushed with death morning, noon, and night."[20] Words like "doughboys" may be outdated now, and phrases like "brushed with death" may appear to be overly dramatic, but one can tell that the respect expressed for forward observers is genuine.

SSgt. Ralph Carver of Company D, 345th Infantry was not alone in his expression of admiration for forward observers and the feeling of security he had knowing that artillery support was always available. Capt. Owen R. O'Neill of the 383rd Infantry Regiment on Okinawa described his forward observers as "heroes all." However, someone with insight and great humility once said: "in any war, the real heroes are the ones who didn't come back." By that strict definition, a significant percentage of forward observers among their fellow artillerymen qualified as "heroes." O'Neill noted that the forward-observation teams "were the greatest single morale factor we had at the front," regardless of whether they were led by an artillery officer or an enlisted man. "The rate of casualties among the officers in charge was so great that replacements many times were impossible. It was then that the NCOs took over. In a great many cases they continued to command sections throughout the remainder of the campaign."[21] Morale was very important, and men like Ralph Carver, Owen O'Neill, and others felt that they and their comrades owed their lives to their effective artillery support.[22] Just as telling were the sentiments of retired Marine Corps general Bernard E. Trainor, who wrote that forward observers had saved his life on more than one occasion.[23]

Historians have asked what motivates men to take part in combat, or in other words, for what do men fight. The answer seems to be not for great causes or patriotic motives, although these may not be discounted entirely, but rather men face up to overwhelming danger in combat because they do not want to let down their comrades. In essence, they fight for each other. U.S. Army training during World War II emphasized the individual importance of each man doing his assigned job as part of a team. When asked why they had performed heroic feats in combat, many soldiers replied, "I was just doing my job."

The general perception of the forward observer was very positive if those around him felt that he was doing his job in a satisfactory manner, though just as negative if he did a poor job. Robert Schroeder, who was a runner for Company G, 410th Infantry Regiment, 103rd Infantry Division, had experience with both good and bad forward observers. One can detect his respect for the one and disdain for the other: "I worked in combat with 2 FOs. One was of the very best and the other the very worst." Lieutenant Manning had the ability to direct fire within a few yards of their positions safely. "The last time I saw him was March 15th '45 sitting beside a road, holding his thigh, trying to stop the blood spurting from a shrapnel wound." The other observer, he declared, was a coward and unwilling to listen to advice from anyone. Schroeder held that man personally responsible for fifteen friendly fire fatalities.[24]

Emblazoned on the walls of the main corridor in Snow Hall, the home of the Field Artillery School at Fort Sill, Oklahoma, are these postwar words of General Patton: "I don't need to tell you who won the war. You know our artillery did."[25] Anyone who has studied military history knows that this is an overstatement. Patton's comments, however, do indicate the tremendous respect he had for the overpowering effectiveness demonstrated by infantry and field artillery working together throughout the war.

Japanese survivors on Bougainville repeatedly attested to U.S. field artillery's deadly effectiveness, obliterating their rifle companies literally to the last man and eliminating their artillery batteries.[26] A German infantryman had this to say about American artillery: "We could see American planes in time to dive into a ditch. We had a chance to hit American tanks with our 88s. But when our positions were smothered without warning by an American artillery concentration—then not even the birds or rabbits could escape. Artillery caused most of our casualties and shell fragment wounds were twice as deadly as bullet wounds."[27]

During the interwar period, many different factors contributed to field artillery's ability to make the quantum leap toward participation in mobile warfare that Scott McMeen and others have described. Certainly motorization, field radios, and the fire-direction center were all important new developments. Yet forward observers in the air and on the ground provided the human element necessary to make the system work, and its performance exceeded almost everyone's expectations. Forward observers on the ground not only did their assigned jobs but also transcended their roles as artillerymen to fight as combat infantrymen, at times providing leadership for the riflemen. Both roles contributed mightily to the war effort.

Epilogue

The forward observer who hung up his binoculars and went home in 1945 would be astounded to see the array of new devices available to determine distance, communicate, and adjust fire today. Excluding weapons and other items carried for survival, the basic equipment he used in that earlier era to call for and adjust fire included a radio or field telephone, a compass, field glasses, and a map or map substitute.[1] Using the coordinates of a known location on the map to establish his position in relation to the firing batteries, the forward observer used his binoculars to visually sense where the initial rounds were landing in relation to the target and then adjusted the fire accordingly. If the batteries had been laid and registered correctly, the accuracy of the fire then depended largely upon the accuracy of both the map and the observer's visual sensing. For a system that depended so much upon the human senses, it was remarkably accurate.

Two technical innovations developed since 1945 have greatly enhanced the forward observer's ability to reduce initial target error: laser range finders and global-positioning systems. Lasers can now be used as target designators to guide projectiles toward their targets, thus reducing the chance of causing collateral damage.[2] Range finding was the first military application of laser technology. U.S. armed forces used the first operational range finders in the mid-1960s. By 1971 a typical field-artillery firebase in the Central Highlands of

Vietnam had among the items at its disposal a pair of 20/40 binoculars, a starlite scope used to see objects in the dark, and a laser range finder.[3]

Laser range finders give the observer the means to determine how far he is from a target or a burst with extreme precision. The use of global positioning, however, has enabled an observer to pinpoint the location of the target almost anywhere on the face of the earth. This enables a forward observer to communicate to a FDC the position of a target in relation to a firing battery, again with an extremely high degree of precision. In 1976 Congress voted to approve funding to develop GPS technology, primarily in concern for national defense, despite its high cost. Two years later the Department of Defense launched the first experimental global-positioning satellite. The Gulf War provided the opportunity to test the system in combat conditions, and it proved highly successful.[4]

Binoculars are still standard equipment today, but now they have a horizontal and vertical reticule patterns or scales divided into given increments in mils. By dividing the horizontal scale into ten-mil increments on both the M-17 and M-19, the observer can measure horizontal deviation from a reference point with a known direction. The M-17 lens has three vertical scales. The forward observer looks at the vertical scales on the left and in the center. These are divided into five-mil increments on the M-17 lens and then used to determine height-of-burst adjustments.[5] One model of binoculars used today combines night vision with a laser range finder, a digital compass, an inclinometer (a device for measuring angles from the horizon), a built-in GPS, and an optional infrared laser pointer. For example, the Leica Vector binoculars currently used by the army are advertised as able to indicate range, azimuth, and inclination.

Today's forward observer also uses a pocket-sized forward-entry device (PFED) for initiating fire missions. This helps target the enemy and sends the request for fire up the chain of command. Slightly larger than a palm pilot, the PFED reduces the potential for miscommunications resulting from weak or garbled radio transmissions or enemy interception. Rugged and durable, the device features call-for-fire icons that distinguish them from other commands. The screen is visible in daylight or darkness, and the PFED operates in all weather conditions, wet, dry, hot, or extreme cold.[6] The World War II–era forward observer would be amazed at this reduction in size and weight of communications equipment. With a battery box weighing twenty-eight and a half pounds and a receiver/transmitter weighing thirty and a half pounds, the old SCR-610 radio used in the 1940s weighed a total of fifty-nine pounds.[7] It is hard to imagine anyone being able to run while under fire with that much weight strapped on his back.

The PFED is actually a miniature personal computer and requires software designed specifically for its use. Currently the U.S. Army uses PSS-SOF as its basic forward-observer software. This gives forward observers the precision capability necessary to engage targets using a variety of exclusive programs as well as a software package used by the U.S. Air Force called joint direct-attack munitions (JDAM).[8]

Another recent change affecting forward observation is the modular reorganization of army combat units from the division to the brigade combat team (BCT). Fires brigades have replaced the division artillery and the BCTs other large-scale unit groupings used for maneuver. A typical brigade includes fewer soldiers than a division. Fires brigades, like any BCT, are not organic or a standard part of any army organization. The intention is to assign each infantry division its own fires brigade as it becomes ready and available for combat.[9] The fires brigade has its own organic units and receives other attached units as required.

December 16, 2004, marked the sixtieth anniversary of the beginning of the Battle of the Bulge. That same day the 4th Infantry became the army's first division to incorporate a fires brigade within its organization.[10] With increased emphasis on the ability to use all available interservice firepower rather than field artillery alone, the fires-brigade concept is intended to make field artillery "the primary executor of army and joint fires for the ground commander in areas not assigned to combat brigade teams (BCTs)."[11] Possibly the greatest consequence of modular reorganization upon forward observation is that it puts more observers in the field at one time, thus decreasing the need to shift them from one area to another as they become casualties while also enhancing the control each has over his area of responsibility.

The switch to the fires-brigade unit means that organic artillery units would now be supporting a smaller overall unit. In addition, the BCT concept allows for increased flexibility in assigning direct-support artillery where needed. In both cases it reduces the number of forward observers necessary to function effectively in combat while making more of them available.

This would seem to indicate that today's process of fire acquisition and control is greatly improved over what is was in 1945—and in most respects it is, though not in all. For example, the time required to engage a target has increased. Maj. Gen. Robert H. Scales, Jr., historian and former commandant of the Army War College, indicated that on average it took about four and a half minutes to initially engage a close-support target and adjust fire in the European theater in 1944. In Korea it took about the same amount of time, but by the time of the Vietnam War, it took about eleven minutes; during the first Gulf War, the delivery of massed fires took an average of fifty-five minutes. Scales attributed the increase in mission time to a variety of factors, including the fear of hitting friendly troops and increased stratification and automation of systems used for air and ground support. Also, today's more expensive munitions generally require more delivery time with the current technology and doctrine.[12]

In the summer of 2001, Maj. Gen. Toney Stricklin, departing chief of field artillery and commandant of the Field Artillery School, observed that among the U.S. Army's most neglected soldiers were fires-support teams and forward observers. He noted that commanders must focus on equipment, training, evaluation, and certification of observers. To make artillery more responsive requires "true sensor-to-shooter linkage," plus streamlining of the digital-communication structure rather than communicating by way of a series of fires-support elements.[13]

In his analysis of the overall effect of post–World War II changes to field artillery, Col. Gary H. Cheek emphasized the need to retain the human element in the fires-support system. While admitting that fires are now more precise and that computerized systems offer more capabilities than previously, he added that it is disturbing to note that although artillery fire is more accurate than ever, the conduct of fire has lost the human dimension and with it some of its responsiveness.[14] Cheek's argument is that under the old system of communication by radio, which enabled the soldiers at the FDC to hear the voice of the forward observer, the urgency and emotion in the observer's voice were an important part of his overall instruc-

U.S. Field Artillery School, Fort Sill, Okla. Photo by author.

tions, important enough perhaps to compel the firing batteries to respond faster than they might otherwise have.[15] He added that to some extent, the new system in which a forward observer requests a fire mission may be likened to posting a query by e-mail.[16]

One organizational change, in addition to the many technological advancements affecting forward observation, is the army's switch to the fire-support-team concept in 1977. With its emphasis upon joint fires and the integration of fires across service branches, the new system designated the enlisted man as the forward observer and a commissioned artillery officer as the fire-support officer.[17] Since then, privates fresh out of their basic training can go to Fort Sill, Oklahoma, where they receive training to become a forward observer. This is quite a change from the 1940s but well justified since enlisted men proved repeatedly during World War II that they were capable of directing artillery fire.

To compare the performance of today's field-artillery forward observers with that of their counterparts in World War II is not the

A forward-observer training class on the firing range, Fort Sill. Photo by author.

Smoke rounds used to bracket a target on the firing range, Fort Sill. Photo by author.

purpose of this study. Aside from all the technological improvements, reorganizations, and new doctrines that have come along
since 1945 is the fact that there are major differences between fighting a limited war and a conventional war. Also, the use of unmanned
aerial vehicles, or drone aircraft, to find targets and direct artillery
strikes has to some extent replaced the activity of humans with that
of machines.

Forward observers today, however, receive recognition for their
service in combat in a way that those in World War II never could.
They are now eligible to wear the Combat Action Badge. As mentioned before, the Combat Infantryman's Badge was created during
World War II as a means of creating esprit de corps among infantryman. Anyone who knows its meaning understands that the soldier wearing it has taken part in combat, and almost from the start,
it became a highly regarded decoration. During the war, forward
observers lived with the combat infantrymen and were actively
engaged in action. But because they were not infantrymen, they
were ineligible to earn the CIB.

During the Vietnam War, tank crews, transportation personnel,
military police, and others with occupation specialties other than
infantry often engaged the enemy in what are defined as combat
conditions. While those trained in armor more routinely took part
in fighting, others who had reason to travel overland frequently
between base camps intermittently took hostile fire and returned
fire. Since that time and more recently, many who have made the
army their career have clamored for a badge similar to the CIB to
indicate their participation in combat. On May 2, 2005, the army's
chief of staff, Gen. Peter J. Schoonover, approved the creation of the
Combat Action Badge (CAB). The badge was retroactive to September 18, 2001.[18]

On June 29, 2005, General Schoonover and Sergeant Major of
the Army Kenneth O. Preston presented the first five CABs in a cer-
emony at the Pentagon. Among these first recipients was a forward
observer, Army National Guard sergeant Timothy Gustafson, who
was wounded while serving in Iraq in January 2004.[19]

Through August 24, 2011, a total of 169 forward observers
serving with the 1st Infantry Division have been awarded the CAB.
Ten more were awarded to men from other units. All but seventeen
of the badges going to forward observers with the Big Red One

were awarded to just three BCTs, or an average of about fifty-one.[20] The product of that average multiplied by the total number of fires brigades that have served in Iraq and Afghanistan since September 2001 would yield a number approximating the total number who have earned that award. Because any solider, regardless of occupational specialty, is eligible to earn the badge, forward observers as a group would seem to be very well represented among all U.S. Army and National Guard personnel having earned the CAB.

The fact that forward observers today are eligible to earn and wear the CAB would undoubtedly please many of those who did that same job during World War II. As mentioned earlier, the presence of forward-observer parties at the front bolstered the confidence of the infantrymen, who were almost universally glad to see the artillerymen among them. At that time officers and enlisted men alike repeatedly recommended forward observers for the CIB. That badge was the envy of all other troops, and although the infantrymen held the award in highest esteem, they were willing to share it with forward observers because they knew these men deserved it.[21] Historians have covered almost every aspect of World War II in depth, though not the story of the forward observers. Few people today even know what the term means. But these frontline artillerymen truly were in the war too.

Notes

Abbreviations

CARL   Combined Arms Research Library, U.S. Army Command and
        General Staff College, Fort Leavenworth, Kans.
MSTL   Morris Swett Technical Library, U.S. Army Field Artillery
        Training Center, Fort Sill, Okla.
NARA   National Archives and Records Administration, College
        Park, Md.
NPRC   Military Personnel Records, National Personnel Records
        Center, St. Louis
RG 407 Records of the Adjutant General's Office, 1917–45, Record
        Group 407

Preface and Acknowledgments

1. Battery A, 334th Field Artillery Battalion, 87th Infantry Division, Morning Report, Feb. 12, 1945, NPRC.

2. Gen. William C. Westmoreland to John R. Walker, May 8, 1996, in author's possession.

3. Dr. Boyd L. Dastrup to John R. Walker, Oct. 9, 1996, in author's possession.

4. Field Manual 3–0, *Operations*, defines combined arms as the simultaneous application of two or more arms, such as infantry and field artillery, to achieve an effect on the enemy that is greater than if they were

used individually in sequence. U.S. Department of the Army, *Operations (DRAG Edition)*, FM 3–0 (Washington, D.C.: Department of the Army, June 15, 2000), 4–27, cited in John W. Washburn, "Integration of Armored Forces in the U.S. Army Infantry Division," 2000, School of Advanced Military Studies Monograph, CARL, PDF, http://cgsc.cdmhost.com/cdm/ref/collection/p4013coll3/id/453, p. 1. Jonathan M. House, the premier U.S. authority on combined-arms warfare, indicates that these tactics represent the application of different combat arms supporting each other on the battlefield. He emphasizes the integration that must take place between the two separate arms. House adds that while professional soldiers have a need to learn such tactics, this is exactly the area where historical documents and field manuals often omit the most significant details. *Toward Combined Arms Warfare*, 3.

5. Lt. Col. James G. Snodgrass defines maneuver in the operational sense as the ability to place combat troops quickly where they may attack an enemy force from almost any angle, cut off his communications, impede his progress, isolate his forces, or escape his attack. Successful maneuver enables a smaller force to defeat a larger one. "Operational Maneuver," 5.

6. Direct fire means that whoever is discharging the weapon observes and takes visual aim at the target. Indirect fire means the shooter cannot see the intended target.

7. Published memoirs of forward observers who served in World War II and after include Edwin V. Westrate, *Forward Observer* (Philadelphia, Pa.: Blakiston, 1944), which might properly be termed "historical fiction"; Jones, *F.O. Forward Observer;* Eugene Maurey, *Forward Observer;* McGhee, *Golden Acorn Memories;* Maki, *Nine Year's Journey;* Major, *Th Memoirs of an Artillery Forward Observer;* Addison Terry, *The Battle for Pusan: A Korean War Memoir* (Novato, Calif.: Presidio, 2000); William H. Hardwick, *Down South: One Tour in Vietnam* (New York: Ballantine Books, 2004); Robert M. Veneable, *Forward Observer* (Ft. Thomas, Ky.: R. M. Venable, 2004); James W. Hengelbrok, *Fire Mission: Observations of a Forward Observer* (Cincinnati, Ohio: J. W. Hengelbrok, 2006); William B. Hanford, *A Dangerous Assignment: An Artillery Forward Observer in World War II* (Mechanicsburg, Pa.: Stackpole, 2008); Franklin Cox, *Lullabies for Lieutenants: Memoir of a Marine Forward Observer in Vietnam, 1965–1966* (Jefferson, N.C.: McFarland, 2010).

8. Dastrup to Walker, Oct. 9, 1996. Howard T. Maki served in World War II as a mortar man, then in Korea as a forward observer. He made the army his career, attaining the rank of lieutenant colonel before retirement. Maki notes: "the artillery f.o. is a tool, albeit an important one, similar to anyone else in the infantry unit to which he is assigned. The difference is that the f.o. lengthens and reinforces the range and capacity of firepower of

the infantry unit and advises the infantry commander in its use. He does not make decisions relating to attack and maneuver; this comes under the purview of the infantry commander. I'm sure this is the reason not much is written or documented about the mission of the forward observer; he is there to assist whenever and however he can help in accomplishment of the mission." Howard T. Maki to John R. Walker, Mar. 7, 2002, in author's possession.

9. Headquarters, Department of the Army, *Table of Organization and Equipment No. 6–327,* 2.

10. House, *Toward Combined Arms Warfare;* House, *Combined Arms Warfare in the Twentieth Century.*

11. Bailey, *Field Artillery and Firepower* (2004).

12. Adkin, *The Charge,* 149.

13. For further reading, see Dastrup, *King of Battle;* Gudmundsson, *On Artillery;* and I. V. Hogg, *History of Artillery.* For further detail about grape shot and canister, see Manucy, *Artillery through the Ages,* 64–69.

14. Nesmith, "Quiet Paradigm Change," 339.

15. Several authors have written about this famed unit. See, for example, Morita, *Nation's Most Decorated Military Unit.*

16. Astor, *Crisis in the Pacific,* xi–xii.

Introduction

1. Remini, *Battle of New Orleans,* 144.

2. John S. D. Eisenhower, quoting Lester R. Dillon, notes: "A canister was a tin cylinder with a powder charge and small shot. Grape consisted of a cluster of balls between two wooden blocks called 'sabots.'" Dillon, *American Artillery in the War with Mexican,* 14, cited in Eisenhower, *So Far from God,* xxiii.

3. For example, Warren Ripley comments on the inconsistency of providing an infantry sniper's rifle, with a range of 500–600 yards with fine telescopic sights, while furnishing the Parrot gun, a standard rifled cannon used by the Union during the Civil War, with a range of 3,000–4,000 yards with sights "far coarser than those of any old smoothbore musket." *Artillery and Ammunition of the Civil War,* 229.

4. J. B. A. Bailey notes that the Russians had used howitzers to experiment with indirect fire as early as the 1750s but that major technical developments involving the use of indirect fire did not appear until late in the nineteenth century. *Field Artillery and Firepower* (2004), 211.

5. Describing that war, Boyd L. Dastrup writes, "cannoneers of the 1890s still stressed closing with the enemy by firing at distances of 800 yards which was well within the range of rifles of the time." *King of Battle,* 140.

6. Vardell Edwards Nesmith, Jr., observes that because of their short duration, neither the Spanish-American War (1898) nor the Second Boer War (1899–1902) in South Africa "challenged the domination of the direct-fire paradigm." But during the Russo-Japanese War (1904–1905), "the lethality of modern weaponry was felt as infantry took to the trenches and the artillery also sought cover. The war tested the direct-fire paradigm for that last time." Nesmith, "Quiet Paradigm Change," 339–40. British historian J. B. A. Bailey writes that "as the First World War approached, direct fire at close range was [yet] the prevailing orthodoxy." *Field Artillery and Firepower* (2004), 206. Bailey explains elsewhere that military leaders worldwide had ignored the fact that technical improvements in small arms gave infantry on the defensive superiority over the tactical mobility of attacking soldiers. In the summer of 1914 that truth became known. Jonathan B. A. Bailey, "The First World War and the Birth of Modern Warfare," in Knox and Murray, *Dynamics of Military Revolution,* 152.

7. Dastrup observes that "even when opportunities existed to employ observed fire, commanders preferred unobserved fire." *King of Battle,* 173.

8. Percin, *Massacre de notre infanterie.*

9. Shrader, *Amicicide,* 1.

10. For an overview of these changes, see Dastrup, *King of Battle,* chap. 7. Professional soldier and historian Col. Robert H. Scales, Jr., has noted, "Since the American Civil War, tactical innovation and change in the American Army have come from the bottom up." *Firepower in Limited War,* 166.

11. Gudmundsson, *On Artillery,* 147.

12. U.S. War Department, *Field Artillery Forward Observation,* 2.

13. General Board, U.S. Forces, European Theater, *Organization and Equipment of Field Artillery Units,* 4; Kerns, *Above the Thunder,* 3–4.

14. John Nelson Rickard has observed that during the Lorraine Campaign, the amount of air support the XIX Tactical Air Command was able to give Patton's Third Army diminished after September due to bad weather and the diversion of planes to other missions. *Patton at Bay,* 83. Edgar Raines writes that "total flying hours [in Europe] dipped appreciably . . . with the onset of autumn rains and the associated fogs in October and did not increase markedly until the weather improved in March, 1945." *Eyes of the Artillery,* 311.

15. Kerns, *Above the Thunder,* 4.

16. Col. Robert H. Scales observes that during the Vietnam War, "the forward observer became the infantry company commander's right-hand man. . . . The survival of the company often depended upon the FO's skill in calling in and adjusting fire quickly and precisely. Good FOs were prized:

bad ones rarely stayed in the field very long." *Firepower in Limited War,* 86. The experiences of forward observers with the 37th and 87th Divisions illustrated in chapters 4–9 will serve to demonstrate that by World War II, this was already the case.

17. After the war infantrymen of the 96th Infantry Division praised their forward observers, noting that during combat the designated observers became casualties so frequently that it was necessary to use NCOs among the parties to command the sections. "Heroes All," 523.

18. A reasonably accurate, though approximate, figure may be possible to compute using all available unit rosters, unit histories, general orders, and other official records but would be extremely difficult to compile. Field artillery incurred 42,692 casualties during World War II. U.S. Adjutant General's Office, *Army Battle Casualties and Nonbattle Deaths in World War II,* 47.

19. Most likely, every American army division that fought in World War II experienced casualties resulting from friendly artillery fire. William Hanford served as a radio operator in the 928th Field Artillery Battalion. He recalled one incident when his own battalion fired at him and the two others with him in his forward-observer group as they went forward to join the infantry. The motor-pool officer directed the fire, thinking he was adjusting smoke shells on enemy machine guns that were holding up an infantry attack. "G Company suffered eight dead and twenty wounded before the sergeant and I could get the radio and battery pack connected and the antenna up to call for a cease fire. Affectively [*sic*] this stopped the entire attack for the day." William Hanford to John R. Walker, Sept. 1, 1999, in author's possession.

20. U.S. War Department, *Field Artillery Forward Observation.*

21. Ralph Carver to John R. Walker, Mar. 13, 2000, in author's possession.

22. Bernard E. Trainor, to John R. Walker, June 1, 1996, in author's possession.

23. Astor, *Crisis in the Pacific,* xi–xii.

Chapter 1

1. Bailey, *Field Artillery and Firepower* (2004), 182–83; House, *Toward Combined Arms Warfare,* 21.

2. House, *Toward Combined Arms Warfare,* 16.

3. John P. Langellier writes that at the time of the Civil War, American smoothbore cannon could reach around 1,500 yards and rifled guns about 2,500 yards. There are 1,760 yards in one mile. He also notes that

the 12-pounder smoothbore "Napoleon" was the mainstay of the Union field artillery. Not as important, but nonetheless technologically significant, were the 3-inch rifled ordnance guns. *Redlegs*, 1, 6.

4. Rifling refers to spiral grooves cut inside of a gun barrel to impart spin to the projectile, giving it greater accuracy over a longer distance. The inside of the barrel of the old smoothbore musket had no rifling, and when fired, the projectile fairly tumbled out of the barrel, causing it to wobble erratically. Peter Paret observes that with a rifled musket, a good shot could hit a target at a distance of 250 yards consistently. *Makers of Modern Strategy*, 419.

5. Scales, *Firepower in Limited War*, 5.

6. S. Jones, "Influence of Horse Supply upon Field Artillery," 357–58.

7. Archer Jones observes that field fortifications were perhaps the most innovative tactic developed during the Civil War, enhancing the power of defense. Grady McWhiney and Perry D. Jamieson note that General Sherman apparently realized earlier than other generals the high price paid for making a tactical assault against entrenched positions, probably during the Atlanta Campaign. A. Jones, *Civil War Command and Strategy*, 38; McWhiney and Jamieson, *Attack and Die*, 106.

8. J. B. A. Bailey notes that shells traveling at high velocity in a flat trajectory cannot hit targets in a depression. *Field Artillery and Firepower* (2004), 209.

9. Prentice G. Morgan, "The Forward Observer." *Military Affairs* 23 (1959): 210.

10. Bailey, *Field Artillery and Firepower* (2004), 196–97.

11. Ibid., 199.

12. Dastrup, *King of Battle*, 96.

13. Garrison, *Friendly Fire in the Civil War*, 79–81.

14. Scales, *Firepower in Limited War*, 8.

15. McKenney, *Organizational History of Field Artillery*, 75–77. Boyd Dastrup defined the breech as "the rear part of a cannon behind the bore." Thus, any breech-loading weapon is one that is loaded near the end of the barrel as opposed to a muzzleloader, where the bullet or shell enters the barrel or bore at the front where it also exits the gun. Dastrup, *King of Battle*, 317. Robert Scales notes: "Gun cotton or nitrocellulose replaced black powder propellants for artillery in the 1880s. Nitro burned more slowly and lessened recoil shock. Less recoil made possible the development of a pneumatic device to halt the rearward movement of a gun and return the tube to the same spot after each round is fired." Gun cotton was also smokeless. Scales, *Firepower in Limited War*, 7.

16. Military historians recognize that military traditions are sometimes hard to change. British historian J. B. A. Bailey observes that the other arms resented artillery's removal to cover and insisted that the guns still be deployed in their sight. He notes how ironic it was that the armies of the world would ignore indirect fire, which would "revolutionize the provision of close support," until World War I forced them into using it. *Field Artillery and Firepower* (1989), 119.

17. Parkhurst, "Artillery at Santiago," 177–79.

18. Shelford Bidwell notes that initially, many officers considered the general use of indirect fire in the years 1914–15 to be a step backward because direct fire was quicker, simpler, and (because it was normally used at shorter ranges than indirect fire) more accurate. Also, if the forward observer was killed or wounded, indirect fire was out of action, at least for a time. Bidwell, "Indirect Fire as a Battle Winner/Loser," in Barnett, *Old Battles and New Defences*, 116.

19. In reference to both Boer Wars, British artillery authority Ian V. Hogg notes that the British had fought using its artillery in the old way of hub-to-hub firing without cover, taking aim at a clearly visible enemy. The Boers, however, hid their field guns and used late-model high-velocity rifles to shoot enemy gunners. From the lessons of this experience came the impetus to use indirect fire and fire from concealed positions. *History of Artillery*, 100. Perhaps not as well known as Hogg, Curt Johnson observes that Boer mounted riflemen posed a real threat to British artillerymen. They were excellent shots and skilled at hitting gunners from afar. As a result, British cannoneers learned to use indirect fire and every bit of concealment they could find in that terrain. In truth, the Boer Wars were the first conflicts in which indirect fire and placing firing batteries in concealment was widely practiced. *Artillery*, 38–39. J. B. A. Bailey observes that the British tried indirect fire during the Boer Wars and "concluded it was impractical in mobile warfare." Bailey, "The First World War and the Birth of Modern Warfare," in Knox and Murray, *Dynamics of Military Revolution*, 136.

20. Nesmith, "Quiet Paradigm Change," 312–14.

21. Johnson, *Artillery*, 43.

22. Nesmith, "Quiet Paradigm Change," 329.

23. Dastrup, *King of Battle*, 158–59. The incorporation of field artillery into divisions was only a provisional measure. Janice McKenney has noted that even by 1912 the U.S. Army was still dealing with the need to create permanent tactical divisional units. *Organizational History of Field Artillery*, 103.

24. Dastrup, *King of Battle*, 153.

25. Snow, "Method of Observing Fire by Use of Lateral Observers," 225.

26. Dastrup, *King of Battle*, 151.

27. John Terraine writes that during World War I, the supremacy of artillery as a destructive force became undisputed. "Indirect Fire as a Battle Winner/Loser," 7. William Odom notes that "World War I witnessed a quantum increase in the destructive power and use of artillery fire." *After the Trenches*, 59. Maj. Gen. William J. Snow later remarked that "this was the first war in history where artillery inflicted more casualties than the small arm or the musket." *Signposts of Experience*, 242. The U.S. Army Medical Corps concluded that artillery caused over 87 percent of all AEF battle casualties. Grotelueschen, *Doctrine under Trial*, 138.

28. Scales, *Firepower in Limited War*, 8.

29. Gudmundsson, *On Artillery*, 69.

30. McGrath, *Fire for Effect*, 16, 32. J. B. A. Bailey observes that the role of U.S. field artillery was limited to simply supporting other arms in a manner of mobile warfare that made the use of indirect fire well-nigh impractical. *Field Artillery and Firepower* (2004), 266.

31. Cooke, *Pershing and His Generals*, 13.

32. Scales, *Firepower in Limited War*, 8. J. B. A. Bailey notes that the firepower of infantry simply ended maneuver, resulting in trench warfare. *Field Artillery and Firepower* (2004), 243.

33. Pierce, "Maximum of Support," 21.

34. Carter, *101st Field Artillery*, 156–57.

35. Grotelueschen, *AEF Way of War*, 165.

36. Gudmundsson, *On Artillery*, 137.

37. Boyd Dastrup writes: "Unable to tie the combat arms into an effective communications network, commanders had to depend upon elaborate plans and rigid schemes of barrages of unobserved fire during the Aisne-Marne offensive and seldom employed observed fire." *King of Battle*, 169.

38. McGrath, *Fire for Effect*, 33.

39. "Extract from Notes on the Methods of Attack and Defense to Meet the Conditions of Modern Warfare," in Combat Studies Institute, "Selected Readings in Military History: Evolution of Combined Arms Warfare," 1983, CARL, 130.

40. Case, "Artillery Support in Attack," 247.

41. House, *Combined Arms Warfare in the Twentieth Century*, 99.

42. Grotelueschen, *Doctrine under Trial*, 139.

43. Shugart, "On the Way," 102–103.

44. Artillery captain David Loring, Jr., describing officers chosen for liaison duty, observes: "During the war he [the liaison] was often the lieu-

tenant who could be most easily spared." Loring, "Instruction in Field Artillery Tactics," 591.

45. Robert Hiner, conversation with author, Aug. 2, 2000.

46. Lanza, "Artillery Support of the Infantry in the AEF," 67.

47. House, *Toward Combined Arms Warfare*, 21–22.

48. Lee, *Artillerymen*, 87.

49. Hays, "Fire Direction of Artillery Supporting Infantry," 510.

50. Shrader, *Amicicide*, 2–3.

51. Dastrup, *King of Battle*, 162.

52. Gudmundsson, *On Artillery*, 38.

53. Augustin M. Prentiss describes the use of artillery to deliver toxic chemicals as "one of the most remarkable developments in the World War." *Chemicals in the War*, 432. Donald Richter notes that because of an ongoing shortage of artillery shells, the British relied heavily on cylinders to deliver gas to enemy lines throughout the war. *Chemical Soldiers*, 36–37. Boyd Dastrup observes that firing gas shells over a broad front could neutralize enemy artillery in two to four hours. *King of Battle*, 162. Mark Groteleuschen notes that gas shells had an advantage over the high-explosive artillery shells normally used when landing among trees. *Doctrine under Trial*, 52.

54. A set-piece attack means that attackers know the general location and strength of the defenders and that the locus of battle is not likely to change once the assault is underway. Mark Grotelueschen observes that after a few years of trench warfare, "most Allied officers also had concluded that the only reasonable course of action was to make limited, meticulously planned, set piece attacks based on crushing artillery barrages." In addition: "There is no record on any training that might have enabled the [AEF] regiments, brigades, or division as a whole to learn how to combine fire and maneuver in some form of mobile attack in an environment devoid of trenches and overwhelming defensive firepower." *AEF Way of War*, 35, 64.

55. Ibid., 104.

56. Hartcup, *Effect of Science on the Second World War*, 45.

57. Stebbins, "Indirect Fire," 76.

58. Grotelueschen, *AEF Way of War*, 358.

59. Ibid., 360.

60. Buchan, *Battle of the Somme*, 243.

61. Dastrup, *King of Battle*, 180–84. David Shugart states, "undoubtedly the Caliber remained Field Artillery's most influential[,] referring to it as the benchmark on what types of future howitzers and guns the field artillery would need." "On the Way," 25.

62. House, *Toward Combined Arms Warfare,* 16.

63. 1919 War Department Annual Report; "Report of the Hero Board: Proceedings of the Board of Offices Convened by the Following Order, General Headquarters, American Expeditionary Forces, Office, Chief of Artillery, 22 March 1919," MSTL, 685 (hereafter Hero Board).

64. Ibid, 686–88.

65. Ibid., 686. Shortly before the attack on Pearl Harbor, Capt. Walter D. Adkins of the field artillery noted that "the majority of those who have practiced the new fire direction technique as taught at the Field Artillery School are more than sold on it. We feel that it is the solution in the effective handling of battalion fires." "This New Fire Direction Technique," 985.

66. Regarding the number of liaison-personnel casualties incurred during World War I, Maj. C. M. Busbee reported: "One Field Artillery Battalion, during the late war is known to have lost 26 men during a period of two days in an effort to keep its telephone lines, principally its liaison lines open. Still another had six liaison officers fall as casualties during a period of a little over a week." L. Jones, "Infantry-Artillery Liaison in Combat," 501.

67. Infantry School, *Infantry in Battle* (Washington, DC: The Infantry Journal, 1939), 256.

68. *Hero Board,* 689.

69. Shugart, "On the Way," 17.

70. Dastrup notes that the Lassiter Board had a major influence on the eventual decision to abandon horse-drawn artillery in favor of motorization. *King of Battle,* 184. From the Westervelt Board came development of the 105-mm howitzer, which Stephen J. Zaloga describes as, "without question, the single most important U.S. Army field artillery piece of World War II." *U.S. Field Artillery of World War II,* 9.

71. Shugart, "On the Way," 19–20.

72. Ibid., 31–32.

73. Ibid., 39–40.

74. Ibid., 62–64.

75. Kirkland, "Orlando Ward," 40.

76. Shugart, "On the Way," 69–70.

77. Fraser-Tytler, *Field Guns in France,* 90–91.

78. Kirkland, "Orlando Ward," 40.

79. Ratliff, "Field Artillery Fire Direction Center," 117.

80. Kirkland, "Orlando Ward," 40.

81. Ratliff, "Field Artillery Fire Direction Center," 118.

82. Kirkland, "Orlando Ward," 40–41.

83. Shugart, "On the Way," 83–84.

84. Blanchard, "Control of the Fire of a Battalion by a Single Forward Observer," 371. Maj. Gen. Robert H. Scales, historian and former commandant of the Army War College, has indicated that on average it took about four and a half minutes to initially engage a close-support target and adjust fire in the European theater in 1944. Scales, "Transforming the Force," 8.

85. McMeen, "Field Artillery Doctrine Development," 37.

86. Kirkland, "Orlando Ward," 41.

87. Shugart, "On the Way," 84.

88. Burns, "Light Artillery Battalion," 351–52.

89. Shugart, "On the Way," 94; Kirkland, "Orlando Ward," 41. Although the Field Artillery School still taught the battery-officer course in 1941, the Officer Candidate School for artillery officers opened in June that year. Sunderland, *Outline History of the FAS,* 102.

90. McGrath, *Fire for Effect,* 59.

91. Jim Carney, "Civil War Tribute to Rise Again," *Akron Beacon Journal,* Mar. 17, 2009, B6.

92. As Robert Scales has observed, "Technology thus favored the defense." *Firepower in Limited War,* 5.

93. Bellamy, *Red God of War,* 28.

94. Scales, *Firepower in Limited War,* 8.

95. Military historian Russell F. Weigley acknowledges that it was the high quality of his communications equipment that enabled a single forward observer, on the ground or more often in a plane, to call for and adjust the fires of all "batteries within range of a target." *Eisenhower's Lieutenants,* 28. Peter R. Mansoor likewise notes that the newly created FDCs and the forward observer equipped with a practical field radio enabled American artillery to impress allies and enemies alike with its ability to mass fires on enemy targets. *GI Offensive in Europe,* 257.

96. Bellamy, *Red God of War,* 30.

Chapter 2

1. In his postwar report, Maj. Gen. Robert S. Beightler wrote: "each campaign saw the Division shoot up greater and greater amounts of artillery ammunition. The figures finally reached astronomical totals, but it is significant that our casualties were proportionately lower, compared to the great amount of fighting we had to do, than almost any other outfit with similar hard fighting." Beightler, *Major General Robert S. Beightler's Report on the Activities of the 37th Infantry Division,* 3 (hereafter Beightler, *Report*).

2. Kedzior, *Evolution and Endurance,* 10–15.

3. Ibid., 16.

4. Shem, "Notes on a Regimental March," 568; Spencer, "Slow March to Military Effectiveness," 91–92.

5. Shem, "Notes on a Regimental March," 568.

6. "Alliance Camp Second in Ratings," *Alliance (Ohio) Review,* Sept. 3, 1935, 1.

7. *37th Infantry Division Pictorial History,* 10.

8. "Centennial of Ohio's 135th Field Artillery," 283.

9. Sligh, *The National Guard and National Defense,* xv.

10. Sligh notes that Executive Order 8530 of September 16, 1940, federalized four National Guard divisions—the 30th, 41st, 44th, and 45th—while ordering coastal artillery and observation units to report to their armories. Ibid., 126.

11. Ohl, *Minuteman,* 75–79.

12. John Ohl notes that, with the threat of immediate U.S. entry into the war apparently less imminent, "many Guardsmen were now eager to get back to their jobs and families. The draftees were only slightly less eager in this regard, even though they were mostly younger than the Guardsmen and had fewer civilian obligations." The intent of the War Department initially had been to limit the length of service to one year. But with the likelihood of reductions in strength as guardsmen and reservists were discharged, in June 1941 it requested Congress to extend the length of military service. In August Congress passed a law obligating all guardsmen and draftees for an additional eighteen months' service. Ibid., 85.

13. Ibid., 86.

14. *Pictorial History of the 37th Division.*

15. Ohl, *Minuteman,* 92.

16. Raines, *Eyes of the Artillery,* 55.

17. Ibid., 19–20.

18. Edgar Raines indicates that most air-corps officers paid little attention to what the two Texans had demonstrated. This may be because the O-47's poor maneuverability rendered it vulnerable to enemy attacks. Another explanation is that the army's emphasis then regarding the air corps was on developing its strategic capabilities, and using planes to direct artillery drew little interest. Ibid., 23.

19. Ibid., 31–32.

20. Wogan, "Air Observation of Field Artillery," 115.

21. Ibid., 116.

22. Raines, *Eyes of the Artillery,* 40. In his memoirs William Wallace Ford buttressed his argument with these points: "Our little flivver plane will have no armament at all. . . . Upon approach of hostile aircraft our pilot will put the little ship into a series of tight turns, barely off the ground;

high speed enemy fighters, much less maneuverable, will have difficulty in bringing their guns to bear." *Wagon Soldier,* 119.

23. Raines, *Eyes of the Artillery,* 80, 66.

24. Ibid., 95.

25. Peek, *Taylorcraft Story,* 147.

26. Raines, *Eyes of the Artillery,* 117–18.

27. Ibid., 288.

28. Peter Mansoor notes that "despite the success of the mobilization system used by the United States during World War II, too many organizations entered combat deficient in the cohesiveness, teamwork, and skills that generally make small units successful in battle." *GI Offensive in Europe,* 48.

29. *Historical and Pictorial Record of the 87th Infantry Division,* 11–12. This reprint edition is a compilation of every individual subunit's history, each being separately paginated.

30. At their annual reunions, surviving members of Battery A, 334th Field Artillery Battalion frequently talked about the relatively small numbers of the original cadre from Camp McCain who were still with the division when it embarked for overseas duty.

31. *Historical and Pictorial Record of the 87th Infantry Division,* 17, 22.

Chapter 3

1. A prime example of disparity in weaponry is the Battle of Omdurman, Sudan, which took place September 2, 1898. Armed with Maxim machine guns, British field marshal Kitchener's outnumbered troops withstood an attack by the fanatical Dervish sect of Baggara Arabs armed only with spears. In the fighting that ensued, the British and their allies lost forty-eight troops, while the spear-toting Dervishes lost 11,000 men. Hallahan, *Misfire,* 285.

2. In December 1941 the editors of *Field Artillery Journal* recognized the contribution of German forward observers to the success of their combined infantry-artillery efforts, noting: "The special importance of observed fire has shown itself clearly. . . . [T]he co-operation between infantry and artillery necessary for the successful support of an attack was always good; and in this case the artillery forward observer played a very important part." "Reasons for the Success of the German Field Artillery," 990.

3. The *Field Artillery Journal* also noted the similarity in field-artillery doctrine between the United States and Germany: "The most striking thing about them is the close similarity with our own texts and doctrines. For years, our artillery commanders . . . have preached mobility, flexibility of command, and fire control, and above all, the ability to concentrate artillery

fire power for maximum effect. The Germans have proved the golden value of these ideals in modern war, and in doing so, have proved our own manuals and methods." Ibid.

4. Jonathan House observes: "During the conquest of Malaya and Burma in 1942, the Japanese made a virtue out of the lack of heavy weapons." *Toward Combined Arms Warfare*, 134. Richard Overy notes "Military culture in Japan demanded the highest sacrifices from soldiers, even a willingness to kill themselves. . . . The unwillingness to surrender, and the ability to survive for long periods of time in conditions of appalling deprivation, made the Japanese soldier a difficult enemy to defeat." *Why the Allies Won*, 222. Allan Millett and Williamson Murray consider tactical air support "rudimentary, partly because of poor communications and partly because pilots were still enthralled with the one-on-one combat of the ancient warrior." They see Japanese artillery support as "generally poor, largely because of quantitative weaknesses, modest firing range, and ammunition shortages." Furthermore, the Japanese army suffered from a lack of unity of command: "The Japanese simply did not have the time to work out many of the practical details of such highly-involved questions as infantry-tank-artillery liaison, control by higher commanders, and logistics of mechanized forces." *Military Effectiveness*, 34–35.

5. U.S. War Department, *Handbook on Japanese Military Forces*, 85.

6. Millett and Murray, *Military Effectiveness*, 34.

7. A 1938 Japanese field manual states that "field or mountain artillery shells passing more than one meter over the heads of friendly troops will not cause any physical damage." U.S. War Department, *Applied Tactics, Japanese Army*, 184.

8. Richard Overy observes that the fighting early on drained the Japanese military of its best resources and that "levels of attrition were too high to build up adequate reserves." *Why the Allies Won*, 223.

9. Field Marshal Slim noted that "the fundamental fault of [IJA] generalship was a lack of moral, as distinct from physical, courage." In other words, Japanese commanders were unwilling to admit that they made mistakes, so they passed on to their underlings the same orders they had received, knowing that with the resources at hand, the mission could not possibly succeed. Millett and Murray, *Military Effectiveness*, 37.

10. McManus, *Deadly Brotherhood*, 172.

11. One of the first books published in the United States during the war dealing with Japanese tactics observes, "Malaya was lost to well-trained soldiers who only had fair equipment but who had in generous measures the human characteristics of will-to-win, stamina, resourcefulness." Thompson, *How the Jap Army Fights*, 118. Allan Millett and Williamson Murray claim: "[I]t [was] the combination of obedience and ferocity that made the

Japanese Army, whatever its condition, so formidable, and which would make any army formidable. It would make a European Army invincible." *Military Effectiveness,* 36.

12. It was the Leibstandarte Division, an element of the Waffen-SS, that was responsible for the infamous Malmedy Massacre of American troops in Belgium on December 17, 1944. Charles Whiting offers two examples to prove that training for members of the Leibstandarte Division may reflect their almost fanatical devotion to duty. Young officer candidates were ordered to dig one-man holes with a shovel. Shortly afterward, a platoon of tanks drove directly over their positions. "It was just too bad for the man who had not dug his hole deep enough." Other training involved exposure to grenade blasts. Instructors ordered each candidate to place a grenade on top of his helmet. After it was balanced there, the instructor withdrew behind cover and ordered the candidate to stand at attention and pull the pin. "Usually there was no damage done if the cadet kept rigidly still and let the explosion dissipate above his steel-clad head. However, if he got rattled and let the grenade fall." *Massacre at Malmedy,* 23.

13. Jonathan House notes that "the fall of France demonstrated not only the importance of combined arms mechanized formations and blitzkrieg penetrations, but also the German advantage over the British and French in combined arms training and procedures." *Toward Combined Arms Warfare,* 86.

14. Richard Overy claims that in just two years the American army became "the most modern army in the world." *Why the Allies Won,* 225. Jonathan House notes that the U.S. Army "gradually corrected these [associated] problems and developed more effective combined arms teams during the breakout from Normandy." *Toward Combined Arms Warfare,* 129.

15. A German artillery officer, Col. M. Blumner, visited the Japanese army in 1937 and reported that "owing to the shortage of horses in Japan, motorization of the artillery is greatly stressed. The heavy artillery is already motorized throughout. At least one fourth of the heavy field artillery [105-mm guns] is motor-drawn. On the other hand, most of the division artillery remains horse-drawn for the present." Merten, "Japan Modernizes Her Artillery," 339. Richard Overy observes: "Japan made no effort to embrace new developments. The army remained an infantry army, reliant largely on horses. The Divisions nominally designated as 'motorised' had only 350 vehicles each." *Why the Allies Won,* 221.

16. Thompson, *How the Jap Army Fights,* 27; Overy, *Why the Allies Won,* 222.

17. Allan Millett and Williamson Murray note that Japanese "armored forces were feeble in tank versus tank (which did not figure in the conflict in China anyhow) and lacked the striking power of western armies." *Military*

Effectiveness, 34. Paul Thompson observes: "The Japanese Army [in 1942] has nothing that can be strictly designated as an antitank gun. . . . [T]he lightest is the 37-mm. model 1922 infantry gun, . . . [which] is yet relatively untried." *How the Jap Army Fights*, 27.

18. U.S. War Department, *Handbook on Japanese Military Forces*, vii.

19. Allan Millett and Williamson Murray mention that other significant shortages included "fuel stocks, medical and veterinary care (including aid stations and casualty clearing centers), road and runway maintenance, bridging and land and sea transportation." On paper the Japanese had a wide range of support services, but "in practice, the quantity of support was scant or nonexistent, and the quality varied from satisfactory to abominable." *Military Effectiveness*, 38.

20. To increase range, the barrel, or tube, of a howitzer must be made thicker to withstand the explosive charges within the breech when the gun is fired. A thicker tube makes a heavier gun. But almost all Japanese guns of a given caliber were around 200 pounds lighter than the equivalent Western guns while the projectile weight was the same. As a result, their projectile velocity and range was less than Western guns. McLean, *Japanese Artillery*, iii.

21. Barker, "Artillery Operations of the New Georgia Campaign," 532.

22. Berman, "Japanese Operations in China," 548.

23. U.S. War Department, *Handbook on Japanese Military Forces*, 93.

24. Wood, "Artillery Notes from Foreign Journals," 317.

25. Johnson, *Artillery*, 90.

26. Col. Milton A. Hill observes that during the Battle of Bataan, "the firing of Japanese artillery was not as good as our own, probably owing to the lack of observation because of our good camouflage." Hill, "Lessons of Bataan," in Thompson, *How the Jap Army Fights*, 132.

27. I. V. Hogg, *German Artillery of World War II*, 8.

28. Cooper, *History of the 110th Field Artillery*, 119.

29. Jonathan House observes that throughout the war, German artillery had to adjust on targets of opportunity using well-known terrain features, and it was extremely difficult and time-consuming for them to mass fires. *Toward Combined Arms Warfare*, 75.

30. Editors, "Reasons for the Success of German Field Artillery," 991. It is interesting too that, despite its effectiveness, Soviet artillery during World War II made little if any use of forward observers, instead relying on much the same system used in World War I. A battery commander, positioned between observers and fire-direction officers, conducted fire. Scales, *Firepower in Limited War*, 162. Chris Bellamy notes that in the Great Patriotic War, the Russians made extensive use of air photography and sound ranging in addition to ground observation, adding that even to

this day, Soviet artillery puts great emphasis on direct fire. *Red God of War*, 162, 215.

31. General Staff, *Tactics of the German Army*, 10.

32. Editors, "A German Reflects upon Artillery," 709.

33. Editors, "Reasons for the Success of German Field Artillery," 990. John Nelson Rickard notes that in 1944, Germany produced 88,000 trucks while losing 100,000 vehicles. In that same year the United States produced 600,000 trucks for the army alone. *Patton at Bay*, 30.

34. Editors, "A German Reflects upon Artillery," 709–10, 714. Curt Johnson claims that using horses to move field guns seemed "archaic" by the 1940s and that many have frequently criticized the Germans for remaining dependent upon horse-drawn artillery. *Artillery*, 89.

35. Gudmundsson, *On Artillery*, 138–39.

36. Headquarters, 37th Infantry Division, "Report After Action: Operations of the 37th Infantry Division, Bougainville, B.S.I., 8 November 1943–1 May 1944," sec. 3, "Artillery Narrative," World War II Operations Reports, RG 407, NARA, 11.

37. Headquarters, 334th Field Artillery Battalion, "History," 80.

38. Scales, *Firepower in Limited War*, 10.

39. Smyth, "Fighting the Nips," 534.

40. Battery C Mothers' Club History, "Monthly Compilation of Letters," [25, 31 32, 43].

41. Frankel, *37th Infantry Division*, 387–97.

42. Michael Clodfelter's study of casualty rates in modern warfare lists total U.S. Army deaths from disease at 26,518 men and indicates that the army suffered 10,828 of these in the Pacific theater. *Warfare and Armed Conflicts*, 960. The Adjutant General's postwar report on casualties confirms Clodfelter's figure of 10,828 for the Pacific theater and gives the corresponding figure for the European theater as 2,474 men. U.S. Adjutant General's Office, *Army Battle Casualties and Nonbattle Deaths in World War II*, 106, 110.

43. U.S. Adjutant General's Office, *Army Battle Casualties and Nonbattle Deaths in World War II*, 8.

44. Headquarters, Americal Division, "Intelligence Annex to Operation Report, Americal Division, Bougainville Operation, 28 December 1943 to 29 April 1944," CARL, 20.

45. Doubler, *Closing with the Enemy*, 240–41.

46. The *Observer's Check List*, compiled for field artillerymen deployed to the Pacific theater, mentions that even flashlight batteries were troublesome. U.S. Field Artillery School, *Observer's Check List, Southwest Pacific Area, 9/26–12/23*, 1942, MSTL, 9, 12.

47. Ferris, "Jungle Communications," 186.

48. F. Norris, "With Mediums," 174–75. Field radios in Europe may have performed well in summer, but cold weather was a different story. The infantrymen of the 5th Armored Division noted that during extremely cold weather, most of their radios froze up. Fifth Armored Division Association, *Paths of Armor: The Fifth Armored Division in World War II* (Nashville, Tenn.: Battery, 1993), 220.

49. Colby, *War from the Ground Up,* 56.

50. David Thibodeau to John R. Walker, Mar. 6, 2001, in author's possession.

51. Most of the historical literature on snipers indicates that artillery observers were high on the list of priority targets in any army. British major H. Hesketh-Pritchard observes that during World War I, snipers could "make it very hot for the enemy's forward artillery observing Officers." *Sniping in France,* 11. Peter R. Senich indicates that during World War II, first on the German sniper's priority list were observers of the enemy's heavy weapons. *German Sniper,* 117. Adrian Gilbert lists the German sniper's standard priority targets as "other snipers, observers, officers and the crews of infantry support weapons." *Sniper,* 89.

52. Describing the experience of the 29th Division with radios in France, Joseph Balkowski notes that "in front-line foxholes, 29th Division riflemen eventually became wary of using their radios because they feared the Germans could pinpoint the source of transmissions by triangulation." *Beyond the Beachhead,* 116.

53. Wells, "Fifteen Days," 231.

54. Casey, "Forward Observer on Guadalcanal," 563.

55. Fuller, "Liaison in the Jungle," 599.

56. Smyth, "Fighting the Nips," 532.

57. Maurey, "Sound Locations," 73; Maurey, *Forward Observer.*

58. Van Horne, "Short-Range Firing against the Siegfried Line," 76.

59. Ostler, "In France . . . with 105s," 170.

60. Gildart, "Artillery on New Georgia," 83–84.

61. Ibid., 86.

62. C.C.O.R. Bulletin, "Lessons Learnt [*sic*] during the Bougainville Campaign," Reference CCOR/55, CARL, 25.

63. F. Norris, "With Mediums," 174.

64. Joseph Balkowski notes that forward observers in Normandy normally sought the highest point available for adjusting fire. In towns, church steeples typically afforded the best places for artillery observation. *Beyond the Beachhead,* 110. Lt. Jim McGhee, a forward observer with the 87th Division, later wrote how he tried to hit a church steeple in the Belgian village of Vesqueville in which German observers were believed to be working. *Golden Acorn Memories,* 35–36.

65. *Cross of Lorraine,* 34–35.

66. Describing the mindset of the great military leaders during World War I, William Hallahan captures the essence of Japanese military doctrine in World War II: "Courage, elan, glory, invincible confidence, pride, battle ribbons, regimental colors, all the manly virtues—these counted for more than any weapon." *Misfire,* 310.

67. Field Artillery School, *Observer's Check List,* 53.

68. David Kaufman to John R. Walker, Sept. 28, 2006, in author's possession.

Chapter 4

1. "History of the Forward Observer," Sept. 12, 1975, MSTL, 1.

2. Boyle, "Has the Close-Support Problem Been Solved?," 385.

3. Frankel, *37th Infantry Division,* 30–31. John Miller indicates that during December 1942, the 37th Division, "the only other complete U.S. Army division in the South Pacific except the American, was then holding the strategically important Fiji Islands and could not be moved." *Guadalcanal,* 214.

4. Stanton, *Order of Battle, U.S. Army, World War II,* 121.

5. Ohl, *Minuteman,* 95–97; Frankel, *37th Infantry Division,* 36–37.

6. Headquarters, 135th Field Artillery Battalion, Vicinity of Tamavua, Viti Levu, Fiji Islands, "Unit History for 1942," Jan. 15, 1943, RG 407, NARA, sheet 2.

7. Frankel, *37th Infantry Division,* 64. In *Chronology, 1941–1945,* compiler Mary H. Williams does not note the arrival of the 37th Division on Guadalcanal on April 6. She indicates that organized resistance on the island ceased on February 9, 1943, but on April 7 a force of seventy-one Japanese bombers and 117 fighters attacked the Guadalcanal area in force, and on September 9 the 37th Division returned from New Georgia to Guadalcanal to train for the upcoming Bougainville operation. Williams, *Chronology,* 91, 103, 132.

8. Casey, "Forward Observer on Guadalcanal," 563. Donald H. Ping, an artilleryman with the 136th Field Artillery Battalion, also commented about the necessity of using high-angle fire. *Smoking War,* 193–94.

9. Charles A. Henne, an officer with the 3rd Battalion, 148th Infantry, has commented that on New Georgia his battalion's patrols carried sound-powered telephones and the SCR-536 radios, but the latter rarely worked when needed. Another radio, the SCR-284, was so heavy and sensitive to the humidity that they routinely left it behind. Henne, *New Georgia Campaign,* 14.

10. Brian Altobello notes that prior to the American invasion of New Georgia, the Japanese removed many of their heavy field pieces to the islands of Enogai, Bairoko, and Vila. Although the Japanese commander on New Georgia had an antiaircraft detachment and a mountain-artillery regiment, the American preinvasion bombardment destroyed many of the guns remaining, and by the time of the landing, many others were without ammunition. *Into the Shadows Furious,* 224.

11. Casey, "Forward Observer on Guadalcanal," 564.

12. Ibid., 566.

13. Ohl, *Minuteman,* 105.

14. Frankel, *37th Infantry Division,* 81.

15. Rodgers, "136th Field Artillery Battalion," 732.

16. Altobello, *Into the Shadows Furious,* 288–89.

17. Rodgers, "136th Field Artillery Battalion," 732.

18. Frankel, *37th Infantry Division,* 83.

19. Stanley Frankel describes fighting on July 28, when the 1st Battalion, 145th Infantry Regiment lost sixteen killed and scores wounded in a single day. Ibid., 96–97.

20. Ohl, *Minuteman,* 112–13.

21. Ibid., 119.

22. From a Letter to Maj. Gen. Robert M. Danford, "Team on New Georgia," 845.

23. Ibid.

24. Janice E. McKenney has observed that shortly after the Spanish-American War, the Ordnance Department appointed "a board of officers to investigate the efficiency of American weapons in the Cuban and Puerto Rican campaigns. The board reported that artillery employment was too limited to produce any useful evaluation." *Organizational History of Field Artillery,* 89. In his account of the Philippine Insurrection, Col. William Thaddeus Sexton notes that while field artillery could be put to good use in the Philippines, it was most effective in an urban environment and situations where the enemy remained in place. Yet it was not well suited for those situations where any degree of mobility was required, noting on several occasions where the infantry could not take along "wheeled transport" of any kind. Sexton notes one occasion when an artillery piece became stuck in the mud and "virtually disappeared in the roadbed." It is also apparent from his writing that the army clung to the use of direct fire during the Philippine War. It is no wonder, then, that military strategists questioned the effectiveness of indirect fire in an jungle environment prior to 1942. *Soldiers in the Sun,* 95, 181, 183.

25. Gildart, "Artillery on New Georgia," 83.

26. Ibid., 87.

27. Headquarters, Division Artillery, "Artillery Narrative, New Georgia," Sept. 22, 1943, APO 37, RG 407, NARA, 2.

28. Gildart, "Artillery on New Georgia," 87.

29. Haines, "Division Artillery in the Battle of New Georgia," 847.

30. Gildart, "Artillery on New Georgia," 87.

31. Hammel, *Munda Trail*, 214.

32. Headquarters, Division Artillery, "Artillery Narrative, New Georgia," 8.

33. Haines, "Division Artillery in the Battle of New Georgia," 847.

34. Headquarters, 37th Infantry Division, General Order No. 40, APO 37, RG 407, NARA, 5.

35. Headquarters, 37th Infantry Division, General Order No. 46, ibid., 4.

36. Headquarters, 37th Infantry Division, General Order No. 38, ibid., 4.

37. U.S. War Department, Circular No. 269, 1.

38. Department of the Army, *History of the Combat Medical Badge*, 52.

39. Astor, *Blood-Dimmed Tide*, 198.

40. Ohl, *Minuteman*, 119.

41. Frankel, *37th Infantry Division*, 107–108.

42. Ohl, *Minuteman*, 120.

43. Beightler, *Report*, 3.

44. Cooke, *Pershing and His Generals*, 142.

45. Close, *Memoirs of a Groundpounder*, 131.

46. Gildart, "Artillery on New Georgia," 88.

47. Ibid.

48. U.S. Army Ground Forces, Field Artillery School, "In the Pacific Areas," Artillery in Combat, 1944–1945, No. 4, CARL, 9.

49. Barker, "Artillery Operations of the New Georgia Campaign," 531.

50. Ibid., 534.

51. Ibid., 536.

52. U.S. Army Ground Forces, Field Artillery School, "In the Pacific Areas," 9.

53. Ibid., 15.

Chapter 5

1. Frankel, *37th Infantry Division*, 119.

2. Gailey, *Bougainville*, 2.

3. Miller, *Cartwheel*, 234.

4. Ohl, *Minuteman*, 126–27; Headquarters, 37th Infantry Division, "Report After Action: Operations of the 37th Infantry Division,

Bougainville, B.S.I., 8 November 1943–1 May 1944," sec. 3, "Artillery Narrative," World War II Operations Reports, RG 407, NARA, 2 (hereafter cited as 37th Infantry Division, Bougainville Artillery Narrative).

5. During the war, American intelligence officials generally rated Japanese artillery as lacking in range, accuracy, and tactical efficiency. McClean, *Japanese Artillery*, iv. The Japanese conduct of surveying, so critical to accurate registration of the guns, was typically simple and inaccurate due to the shortage of personnel. U.S. Army Ground Forces, Field Artillery School, "In the Pacific Areas," Artillery in Combat, 1944–45, No. 4, CARL, 81.

6. Gailey, *Bougainville*, 108–109.

7. Headquarters, 135th Field Artillery Battalion, "Battalion History, April–December 1943: New Georgia Island, Solomon Islands Campaign," Jan. 6, 1944, RG 407, NARA, 3.

8. During Major Fagan's raid, two U.S. destroyers heavily shelled the coast as Capt. Milton B. Bagby of the 136th Field Artillery observed. 37th Infantry Division, Bougainville Artillery Narrative, 4.

9. Professional artillerymen David P. Valcourt, Robert T. Bray, and Tommy A. Williamson observe that today, U.S. field artillery "is in the process of solidifying the FA's role as the Army's joint fires and effects integrator." "2004: The State of Field Artillery," 2.

10. Shaw and Kane, *Isolation of Rabaul*, 270–71.

11. 37th Infantry Division, Bougainville Artillery Narrative, 3–4.

12. Ibid., 4.

13. Guenther, "Artillery in the Bougainville Campaign," 333.

14. Shaw and Kane, *Isolation of Rabaul*, 291.

15. Ohl, *Minuteman*, 128–29.

16. Miller, *Cartwheel*, 352.

17. Ohl, *Minuteman*, 128–29.

18. Beightler, *Report*, 5; Frankel, *37th Infantry Division*, 142; Ohl, *Minuteman*, 129. For details of the 6th Division's involvement in atrocities in China, see Chang, *Rape of Nanking*.

19. Gailey, *Bougainville*, 136–37. Stanley Frankel records the same incident but with no mention of friendly fatalities, indicating in his account that during the night, the reinforcements set up a new perimeter several hundred yards to the rear. During the artillery barrage the next morning, several short rounds fell into the area, wounding fourteen members of the 129th Infantry Regiment. *37th Infantry Division*, 139.

20. In 1945 the army defined dispersion errors as "errors inherent in the dispersion pattern (such as those caused by manufacturers' tolerances and those errors inherent in the piece and the ammunition)." The conclusion was that such errors could not be controlled, only errors due to

personnel could be reduced. U.S. War Department, *Field Artillery Gunnery*, 16. Despite concern that the extensive use of barrels or tubes could cause bursting, the field artillery placed a heavy emphasis on the importance of "keeping ammunition lots straight . . . and the avoidance of mixing zone weights of shell without appropriate correction. The shooting of one and only one ammunition lot is of utmost importance in the 105mm. How. [*sic*] and smaller calibers." General Board, U.S. Forces, European Theater, *Condemnation and Replacement of Artillery Tubes in Combat*, 2.

21. Nichols, *Ernie's War*, 180.

22. Cocklin, "Bougainville—1944," 452.

23. Miller, *Cartwheel*, 354.

24. Ibid., 356–57.

25. Gailey, *Bougainville*, 150. John Miller notes: "Hill 700 which commanded the entire beachhead was steep, with slopes of 65 to 75 percent in all directions. American intelligence estimates, though not ruling out an enemy attack here, had tended to discount its probability." *Cartwheel*, 359.

26. Miller, *Cartwheel*, 364.

27. 37th Infantry Division, Bougainville Artillery Narrative, 9.

28. Headquarters, 37th Infantry Division, General Order No. 43, Apr. 28, 1944, APO 37, RG 407, NARA, 3.

29. Frankel, *37th Infantry Division*, 144.

30. Ibid., 145.

31. Ibid., 146.

32. Headquarters, 135th Field Artillery Battalion, "Battalion History, April–December 1943: New Georgia Island, Solomon Islands Campaign, Part Two," RG 407, NARA, 3.

33. Ibid., 4.

34. Guenther, "Artillery in the Bougainville Campaign," 334.

35. 37th Infantry Division, Bougainville Artillery Narrative, 11.

36. Frankel, *37th Infantry Division*, 166. John Miller does not describe this incident as such but notes that "the Americans burned, dug, and blasted the Japanese out of their ravines, trenches, foxholes, and pillboxes while the seven artillery battalions, their fire directed by General Kreber, . . . shelled the concentrated enemy troops in front of the American lines." *Cartwheel*, 376–77.

37. Guenther, "Artillery in the Bougainville Campaign," 334.

38. 37th Infantry Division, Bougainville Artillery Narrative, 11.

39. Frankel, *37th Infantry Division*, 166.

40. Headquarters, 135th Field Artillery Battalion, "Battalion History: History of the 135th Battle Campaign on Bougainville, Part Two," 11.

41. Headquarters, 37th Infantry Division, General Order No. 46, APO 37, RG 407, NARA, 1; General Order No. 36, ibid., 2.

42. Headquarters, 37th Infantry Division, General Order No. 39, ibid., 1–2.

43. Headquarters, 37th Infantry Division, General Order No. 48, ibid., 4.

44. Beightler, *Report*, 5. Harry Gailey places the number of Japanese on Bougainville prior to the American invasion at 65,000 men. He estimates that the Australians killed 18,000 of the enemy after U.S. forces departed and that 21,000 survived until the surrender. This would put the total number of Japanese killed by all American units at about 26,000. *Bougainville*, 211. John Ohl states that the Japanese lost 5,000 dead and 3,000 wounded in Operation TA alone. *Minuteman*, 138.

45. Beightler, *Report*, 5. The 37th Division's report for Bougainville indicates that from November 8, 1943, through May 1, 1944, the division fired a total of 161,968 rounds. Each campaign saw the 37th use more artillery fire. 37th Infantry Division, Bougainville Artillery Narrative, 14.

46. Gailey, *Bougainville*, 186.

47. Frankel, *37th Infantry Division*, 145.

48. Most historians attribute this phrase to North Vietnamese lieutenant colonel Nguyen Huu An, who said, "Move inside the column, grab them by the belt, and thus avoid casualties from the artillery and air." As quoted in Lt. Gen. Harold L. Moore, USA (Ret.), and Joseph Galloway, *We Were Soldiers Once and Young: Ia Drang, the Battle that Changed the War in Vietnam* (New York: Random House, 1992), 230.

49. The *Handbook on Japanese Military Forces* explains that "Japanese tactical doctrine insists vigorously on the inherent superiority of the offense," adding that "the object of all maneuver is to close quickly with the enemy." The Japanese somehow view defensive combat as a negative form of action. War Department, *Handbook on Japanese Military Forces*, 85–86.

50. Scales, *Firepower in Limited War*, 86.

51. Shaw and Kane, *Isolation of Rabaul*, 291.

52. Guenther, "Artillery in the Bougainville Campaign," 330.

53. Yourada Toigoichi, 23rd Infantry: "The artillery fire of 9 March killed many men and last night's barrage (13 March) nearly killed all." Yaewake Akiyoshi, 23rd Infantry: "Our artillery was practically all knocked out." Kawakami Kazui, 23rd Infantry: "Half of the 10th Company was wiped out by U.S. artillery before it reached the wire." Nagatomo Milsukuni, 23rd Infantry: "The 8th Company was practically annihilated by U.S. artillery fire on Temmoku Mountain on 13 March." Shimobukoro Yashio, 45th Infantry: "The U.S. artillery barrage of last night (24 March) virtually annihilated the Butai and remnants are scattered. The barrages hit us before the time of our proposed attack." Sakamoto Mosao, 6th Field Artillery: "The second battery numbered about sixty men. When it was at Kuji

Mountain there were 130 men but this was reduced by one half by U.S. artillery fire and after yesterday's barrage, only thirty are left. The battery's two guns were knocked out by U.S. artillery." Yamashita Tatsup, 6th Field Artillery: "The regiment brought eighteen guns to the sector but nearly all have been knocked out by U.S. artillery fire. The strength of the 1st Battery was about 150 men but this was reduced to twenty by artillery fire and after last night's barrage (24 March) not a single man was left." 37th Infantry Division, Bougainville Artillery Narrative, 15.

Chapter 6

1. Rogers, "Cagayan Valley Operation," 100.
2. Frankel, *37th Infantry Division,* 296.
3. Smith, *Triumph in the Philippines,* 4.
4. Ibid., 16.
5. Ohl, *Minuteman,* 155.
6. Headquarters, 37th Infantry Division, "Report After Action: Operations of the 37th Infantry Division, Luzon, P.I., 1 November 1944– 30 June 1945 (M-1 Operation)," pt. 6, "Artillery," CARL, 259 (hereafter 37th Infantry Division, Report After Action: Luzon).
7. Frankel, *37th Infantry Division,* 229–30.
8. Ohl, *Minuteman,* 161.
9. 37th Infantry Division, Report After Action: Luzon, 260.
10. Hearn, "Early Luzon Experience," 328.
11. Rogers, "Cagayan Valley Operation," 101.
12. Sackton, "Battle Notes of Division Artillery on Luzon," 539.
13. Frankel, *37th Infantry Division,* 234.
14. Headquarters, 37th Infantry Division, General Order No. 99, APO 37, RG 407, NARA, 3.
15. Les Boren, interview by John R. Walker, Alliance, Ohio, Aug. 26, 2006.
16. Headquarters, 37th Infantry Division, General Order No. 88, APO 37, RG 407, NARA, 2.
17. 37th Infantry Division, Report After Action: Luzon, 260.
18. Ohl, *Minuteman,* 165.
19. Frankel, *37th Infantry Division,* 239.
20. 37th Infantry Division, Report After Action: Luzon, 261.
21. Headquarters, 37th Infantry Division, General Order No. 68, APO 37, RG 407, NARA, 2.
22. Ohl, *Minuteman,* 170.
23. Smith, *Triumph in the Philippines,* 249–50.
24. Frankel, *37th Infantry Division,* 260.

25. The division's after-action report notes that "this was the first time the Japs had used massed fires of heavy weapons on such a heavy scale." 37th Infantry Division, Report After Action: Luzon, 262.

26. Ohl, *Minuteman*, 180.

27. Robert Ross Smith observes that "the losses had manifestly been too heavy for the gains achieved." *Triumph in the Philippines*, 264.

28. Ohl, *Minuteman*, 182.

29. Ibid., 183.

30. Ibid., 184.

31. Ibid., 185. Although there are several different types of ammunition available for the same caliber weapon depending upon the type of target, the standard HE (high explosive) projectile for an American 105-mm howitzer during World War II was a streamlined shell weighing 33 pounds, filled with 4.8 pounds of TNT. Ian V. Hogg, *British and American Artillery of World War II* (Mechanicsburg, Pa.: Stackpole, 2002), 62. J. B. A. Bailey indicates that there is an inverse relationship between the destructive power of a single shell and the area it can damage. *Field Artillery and Firepower* (2004), 7.

32. Ohl, *Minuteman*, 187.

33. Frankel, *37th Infantry Division*, 269.

34. Ohl, *Minuteman*, 190–91; Smith, *Triumph in the Philippines*, 294.

35. Supreme Commander of Allied Powers, *Reports of General MacArthur*, 275.

36. Ohl, *Minuteman*, 191.

37. Frankel, *37th Infantry Division*, 288.

38. Capt. Glenn A. Steckel indicates that the purpose of the direct fire was to expand existing gaps in the wall and to open new gaps to provide additional openings for the infantry assault that would follow. The 37th Division's artillery would use indirect fire "to reduce and destroy obstacles such as mine fields and barricades in the immediate path of the assaulting troops." "Role of Field Artillery in the Siege on Intramuros," 5; The division's after-action report notes that direct fire was used on the walls, entrances, and areas of the building around the entrances. 37th Infantry Division, Report After Action: Luzon, 262. Another explanation is that at Intramuros, the 37th Division used direct fire to provide destructive force that typically requires a much greater expenditure of ammunition. Its batteries fired indirectly to achieve neutralization that has the purpose of "causing severe losses, hampering or interrupting movement or action, and in general destroying the combat efficiency of enemy personnel." U.S. War Department, *Field Artillery Field Manual: Tactics and Techniques*, 97.

39. Garay, "Breach of Intramuros," 8–10.

40. Steckel, "Role of Field Artillery in the Siege on Intramuros," 8.

41. Smith, *Triumph in the Philippines*, 296.

42. Steckel, "Role of Field Artillery in the Siege on Intramuros," 10.

43. Garay, "Breach of Intramuros," 10–11.

44. Frankel, *37th Infantry Division*, 288–89.

45. Ohl, *Minuteman*, 192.

46. Ibid., 192–93; Smith, *Triumph in the Philippines*, 299. Very few adult males were among those released. Robert Ross Smith states only that the Japanese had murdered them. Ibid, 299–300. Stanley Frankel notes: "Few men were observed as the Japs had removed the males to Fort Santiago and burned them in a room twenty-five feet square where they were later found dead, five layers deep." *37th Infantry Division*, 291.

47. Ohl, *Minuteman*, 193.

48. Frankel, *37th Infantry Division*, 292–93.

49. Smith, *Triumph in the Philippines*, 303.

50. Frankel, *37th Infantry Division*, 293–94.

51. Rodgers, "136th Field Artillery Battalion," 733.

52. J. Gordon, "Battle in the Streets," 28–29.

53. Smith, *Triumph in the Philippines*, 306.

54. Wright, *1st Cavalry Division*, 134.

55. Frankel, *37th Infantry Division*, 303, 307.

56. Ibid., 308–309. The division's after-action report indicates that on April 9 the 129th Combat Team received heavy artillery fire near Salat and that Lieutenant Colonel Brown died "while directing counter-battery fire from a forward OP." 37th Infantry Division, Report After Action: Luzon, 263.

57. Ohl, *Minuteman*, 202–203.

58. 37th Infantry Division, Report After Action: Luzon, 263.

59. Frankel, *37th Infantry Division*, 316.

60. Ohl, *Minuteman*, 204.

61. 37th Infantry Division, Report After Action: Luzon, 264.

62. Ohl, *Minuteman*, 204.

63. Ibid., 206. Stanley Frankel notes: "In New Georgia, Bougainville, and Manila, air support was either non-existent or patently ineffective. The officers and men had no confidence in the 'up close' technique." *37th Infantry Division*, 319.

64. Headquarters, 37th Infantry Division, General Order No. 150, APO 37, RG 407, NARA, 4.

65. Smith, *Triumph in the Philippines*, 561.

66. Frankel, *37th Infantry Division*, 331.

67. Ibid., 332.

68. Ibid., 334.

69. Ibid., 339–40.

70. 37th Infantry Division, Report After Action: Luzon, 266.

71. Rogers, "Cagayan Valley Operation," 101.

72. Frankel, *37th Infantry Division*, 342–43.

73. Ibid., 345.

74. Headquarters, 37th Infantry Division, General Order No. 295, APO 37, RG 407, NARA, 1.

75. Frankel, *37th Infantry Division*, 351.

76. Ohl, *Minuteman*, 217.

77. Frankel, *37th Infantry Division*, 345. After the war historian Robert Ross Smith wrote a letter to Beightler, noting that the Cagayan Valley operation "was a prime example of how much chaos and confusion a relatively light force can achieve when it keeps going as fast and as hard as it can." Ohl, *Minuteman*, 217.

78. In his postwar report General Beightler mentions that at times the front and rear elements of the division were separated by as much as fifty miles. Beightler, *Report*, 14. One newspaper correspondent noted, "The 37th Division, racing up the Cagayan Valley, has a front line two hundred miles long and twenty yards wide." Frankel, *37th Infantry Division*, 345.

79. Ibid., 340, 348.

80. Rogers, "Cagayan Valley Operation," 102.

81. Ohl, *Minuteman*, 220–21.

82. Beightler, *Report*, 14.

83. Smith, *Triumph in the Philippines*, 307.

84. British authors Richard Connaughton, John Pimlott, and Duncan Anderson conclude that "the fault lies with commanders who . . . once presented with the need to take the city by force, preferred to solve the problem with firepower." *Battle for Manila*, 200. Filipino author Alfonso J. Aluit believes that "Douglas MacArthur bears as much responsibility as Sanji Iwabuchi does for the cruel fate that was inflicted on Manila." *By Sword and Fire*, 395.

85. Robert Ross Smith submits that although neither the United States nor Japan had signed the Geneva Convention, both nations had agreed to abide by its rules. Japan forsook this agreement in Manila by fortifying hospital buildings, "all clearly marked by large red crosses on their roofs and they contained many Filipinos, who were in effect, held hostage by the Japanese." *Triumph in the Philippines*, 286.

Chapter 7

1. Stanton, *Order of Battle, U.S. Army, World War II*, 159.

2. Few historians would argue that the National Guard units were particularly well equipped in December 1941, but the point here is that many field artillerymen who served in the guard had comparatively more

training and cohesion than their regular-army counterparts who joined their units after the declaration of war. Peter Mansoor has noted: "too often companies and battalions were forced to adept on the battlefield to achieve the level of combat effectiveness that ideally should have come from progression through the Mobilization Training Program." *GI Offensive in Europe*, 48. Russell Weigley underscores this point indirectly, observing: "Artillerymen largely do the same thing in combat that they have done through all their training—laying down fire on targets they do not see." *Eisenhower's Lieutenants*, 28. Some prewar guard artillery batteries maintained much of their unit members' identity through the entire war. For a time the guard's recruiting slogan was "Join the Guard and go with the boys you know." W. D. McGlasson, "Mobilization 1940: The Big One," *National Guard* (Sept. 1940): 10, as cited in Sligh, *The National Guard and National Defense*, 125.

3. Headquarters, 345th Infantry Regiment, "History of the 345th, 1 December 1944 thru 31 December 1944," RG 407, NARA, 3–6.

4. *Fifth Infantry Division in the ETO*, 131.

5. For further reading on the Battle of Metz, see ibid.; Kemp, *Unknown Battle;* and Gajdusek, *Resurrection.*

6. Headquarters, 334th Field Artillery Battalion, Unit History, Dec. 1944, RG 407, NARA, 2.

7. Sam Delli, conversation with author, Wadsworth, Ohio, May 5, 2007.

8. "History of the 346th Infantry," in *Historical and Pictorial Record of the 87th Infantry Division*, 61. Bruce Egger, coauthor of *G Company's War: Two Personal Accounts of the Campaigns in Europe, 1944–45,* notes that elements of the 87th Division relieved the Yankee Division's 2nd Battalion, 101st Infantry Regiment near Obergailbach on about the same date. He could tell the 87th was a green division because the men practiced very poor noise discipline as they moved into the line, talking out loud and rattling their mess kits. Bruce E. Egger to John R. Walker, Jan. 17, 1995, in author's possession.

9. Williams, *Chronology*, 352.

10. "Historical Record of the 912th FA Battalion," in *Historical and Pictorial Record of the 87th Infantry Division*, 155.

11. "Invictus on the March: The 345th Infantry in Action in the European Theater of Operation," in *Historical and Pictorial Record of the 87th Infantry Division*, 65.

12. Ehret and Granton, *As if on Automatic Pilot*, 27.

13. Dr. Charles F. Ehret, e-mail message to author, Jan. 20, 1999, in author's possession.

14. Ironically, the National Archives holds unit histories of the 87th Reconnaissance Troop for January–May 1945, but the December 1944

record is missing. 87th Cavalry Recon Troop, Unit History and After Action Reports, Box 12921, RG 407, NARA.

15. Ehret to Walker, Mar. 25, 1999.

16. Sam Jones, "To Hell and Almost Back," [1990], Sam Jones Papers, U.S. Army Heritage and Education Center, U.S. Military History Institute, Carlisle, Pa., 13.

17. McGhee, *Golden Acorn Memories,* 16.

18. Regarding American field radios, U.S. Army historian Hugh M. Cole observes that the "radio of the type used in 1944, lacked the necessary range and constantly failed in the woods and defiles." Walker was carrying his unit while in a deep ravine with thick woods at the top. *Ardennes,* 654.

19. Gayle Bricker recalled McGhee leaving them in the rock pile, adding that as a nineteen-year-old, he remembered lying there scared out of his wits. Don Walker, meanwhile, took off his pack " and was leaning back on it chewing gum like it was Sunday afternoon. We were shelled for about four hours." Gayle A. Bricker, Jr., interview by John R. Walker, in Cuyahoga Falls, Ohio, Sept. 30, 1995.

20. Paul Cutler to John R. Walker, June 3, 1999, in author's possession.

21. Ehret to Walker, Jan. 20, 1999.

22. Cutler to Walker, June 3, 1999.

23. Donald F. Hoehle to John R. Walker, Apr. 3, 1998, in author's possession.

24. Hoehle, "I'll Always Be Grateful to Them."

25. Donald L. Walker to John R. Walker, Dec. 18, 1966, in author's possession.

26. Headquarters, 87th Division, General Order No. 245, RG 407, NARA, 1.

27. Command Post, 87th Division Artillery, S-2 Record of Enemy Activity, Dec. 16, 1944, RG 407, NARA.

28. Headquarters, XII Corps, S-2 Record of Enemy Activity: 87th Infantry Division Breakdown, Dec. 13, 1944, NARA, 1.

29. Jim McGhee sincerely believed that Company L had been reduced to some "thirty effectives." *Golden Acorn Memories,* 19. Nevertheless, Company L morning reports for December 1944 do not seem to bear this out.

30. Guy E. Allee to John R. Walker, Feb. 6, 2000.

31. Williams, *Chronology,* 354.

32. *Combat History of the 44th Infantry Division,* 24.

33. Ibid.

34. *History of the United States Twelfth Armored Division,* 29.

35. "Liaison officer from the 12th Armored Division reports that their G-2 has civilian info [*sic*] that valley E of the town of Medelsheim and

Seyweiler is a trap. Reports also state that two towns mentioned above have heavy artillery positions located within and near them. Claim that advance up the valley would be disastrous in that both ridges flanking valley have well located positions." This form was signed and dated at 11:45 A.M. on December 16. 87th Division Artillery, "Message to C, S, G-3 Ln, 16 Dec 44," S-2 Record of Enemy Activity, Dec. 16, 1944.

36. Company L, 345th Infantry Regiment, 87th Division, Morning Reports, Dec. 1944, NPRC.

37. "Invictus on the March," 171–72.

38. U.S. Army, "Casualties, Saar Valley, 16 December 1944," in "Alphabetical Listing: Battle Deaths of the 87th Infantry Division by Organization as of June 1947," CFN 161, RG 407, NARA.

39. Ridings, "Friendly Fire," 31–32.

40. John E. Long to John R. Walker, Feb. 26, 2000, in author's possession.

41. 87th Infantry Division, "345th Unit History—December, 1944," RG 407, NARA, 4.

42. Shrader, *Amicicide*, 12.

43. Rusiecki, *Key to the Bulge*, 25–32. Although every nation's armed forces suffered some friendly fire casualties during World War II, it appears likely that more Americans became casualties as a result of friendly supporting fires in Vietnam. By the 1960s, airstrikes augmented artillery fire much more frequently. Col. David Hackworth, a Vietnam veteran, estimated that 15–20 percent of all U.S. casualties were caused by friendly fire. Appy, *Working Class War,* 185.

44. Editors, "A German Reflects upon Artillery," 714. Military historian Hugh M. Cole later concurred with the general's assessment: "There is considerable indication that German commanders quickly recognized and made gainful tactical use of the gaps at division and corps boundaries where the defenders conventionally failed to provide overlapping fire between zones." *Ardennes,* 659.

45. Addams, *Best War Ever,* 79.

46. Van Creveld, *Fighting Power,* 33.

47. Headquarters, 87th Division Artillery, After Action against the Enemy Report, (010600 December 1944 to 010600 January 1945), RG 407, NARA, 1.

Chapter 8

1. Regarding the Battle of the Bulge, Hugh Cole has observed: "There is no doubt that German artillery helped the assault waves forward during the rupture of the American forward defensive positions. It is

equally clear that German artillery failed to keep pace with the subsequent advance." *Ardennes,* 657. He also notes that by December 23, the Americans had brought a total of 4,155 artillery pieces into action. This would have been before most units of Patton's Third Army had entered the battle. Ibid., 659.

2. Ibid, 657. American counterbattery fire may have been at a disadvantage while the weather was bad for flying. Hugo Gisske of Los Altos, California, was a captain with Headquarters Battery, 912th Field Artillery, served as a liaison officer, and was awarded the Bronze Star and Purple Heart. Gisske noted that the forward observer on the ground seldom engaged in counterbattery fire. This makes sense because to be close enough to see the enemy batteries would most likely mean that he was behind enemy lines. In any case, Gisske explained that the forward observer's range of targets was confined to his immediate front. Hugo Gisske to John R. Walker, Oct. 9, 2001, in author's possession.

3. Peter Elstob notes that Gen. Hasso von Manteuffel's Fifth Panzer Army fought in the Ardennes Offensive after having already been in continual action since leaving the Eastern Front. *Hitler's Last Offensive,* 31. Hugh Cole observes that at the opening of the offensive, "it is probable that the over-all ratio of Germany infantry to American was three-to-one with a ratio of six-to-one at points of concentration." *Ardennes,* 650.

4. Cole, *Ardennes,* 658.

5. Ibid., 78.

6. Cavanaugh, *Krinkelt-Rocherath,* 57.

7. Raymond R. Jemc, "Forty Years Ago," Jemc Papers, n.d., U.S. Army Military History Institute, Carlisle Barracks, Pa., 9–10.

8. McGhee, *Golden Acorn Memories,* 35.

9. Foreman, "Other Side of the Hill," 21.

10. McGhee, *Golden Acorn Memories,* 36–37.

11. Rowland J. Coonradt, conversation with author at Battery A annual reunion, Sept. 6, 2008.

12. Headquarters, 334th Field Artillery Battalion, Unit History, Mar. 1945, RG 407, NARA, 3.

13. Cole, *Ardennes,* 643.

14. Elstob, *Hitler's Last Offensive,* 357.

15. Robert Cruzan, "The Bulge," in Charles R. Foreman, comp., "The Bulge Battle of the Ardennes, First Battalion 345th" (unpublished battalion history, 345th Infantry Regiment, 87th Infantry Division, Marquette, Mich., n.d.), 165–66.

16. Ibid., 167–68.

17. Ibid, 185.

18. Everett R. Criss, e-mail message to author, Mar. 20, 2000, in author's possession.

19. Robert Booth has observed that any attack on the village had to originate across a vast, open area, thus placing attacking infantrymen in an extremely vulnerable position. There were stone buildings along both sides of the street in the village, offering defending troops excellent protection. The ground rising behind Bonnerue gave German tanks optimal placement for lending their support. Robert T. Booth to John R. Walker, Mar. 7, 2000, in author's possession.

20. "Invictus on the March: The 345th Infantry in Action in the European Theater of Operation," in *Historical and Pictorial Record of the 87th Infantry Division*, 71.

21. Booth to Walker, Mar. 7, 2000.

22. Ibid.

23. Foreman, "The Bulge," 185.

24. Headquarters, 87th Infantry Division, General Order No. 88, RG 407, NARA, 1; General Order No. 244, ibid., 1.

25. General Board, U.S. Forces, European Theater, *Organization and Equipment of Field Artillery Units*, 4.

26. Hugh Cole observes that the 2nd Infantry Division suffered unusually high losses of forward observers: thirty-two out of forty-eight working within a single field-artillery battalion required evacuation for wounds or exposure within the first week of the battle. *Ardennes*, 125.

27. Army records from World War II rarely identify individuals as forward observers in any capacity and even less so as casualties. As indicated previously, the history of the 912th Field Artillery Battalion indicates that a forward observer became the first casualty in the battalion, and another forward observer and his radio operator were killed the next day. "Historical Record of the 912th FA Battalion," in *Historical and Pictorial Record of the 87th Infantry Division*, 155. The personal papers of men like Ray Jemc, who served on forward-observation teams, provide some of the fullest details about the casualties incurred among these men. Over the years, division association newsletters, including the 87th Division Association's *Golden Acorn News*, also have provided many historical facts and details that might have otherwise been lost forever. Years after the war, Ray Jemc told his son, Robert, that among the thirteen other individuals in the battalion who shared forward-observation duty with him in combat operations, he was the only one who was not physically wounded or emotionally scarred by the experience. Robert T. Jemc, telephone conversation with author, Mar. 11, 2012.

28. Headquarters, 87th Infantry Division, General Order No. 244, RG 407, NARA, 1–2.

29. Headquarters, 87th Infantry Division, General Order No. 4, ibid., 2.

30. Pattison as cited in Cole, *Ardennes,* vii. Lt. Col. George Ruhlen, who commanded an American field-artillery battalion during the Ardennes Offensive, observed that "the German attack timetable was fatally disrupted in the first four days of the Bulge, in numerous cases by squads, platoons, companies—and sometimes individual soldiers—who held onto key terrain until killed, captured, or forced to withdraw." Astor, *Blood-Dimmed Tide;* ix–x. John Toland describes the Battle of the Bulge as "unorthodox; divisions, regiments, battalions, companies, at times, even one or two men fought lonely battles that determined great issues. In this kind of fight, the American soldier excelled. . . . [Many Germans] still firmly believe that what beat them was an overwhelming number of bombs and shells. . . . But the GI never cared about the chivalry of war. He wanted only to win and go home." *Battle,* 379–80.

31. Headquarters, 87th Infantry Division, General Order No. 9, RG 407, NARA, 9, 12. Whether Hildebrand and Jackson later received direct commissions is unknown. An official study notes that enlisted artillerymen who received direct commissions "would have performed their missions much more efficiently if they had had a better knowledge and training in gunnery." While there is no indication here that the pair actually hit the enemy tank, they did succeed in forcing its withdrawal. General Board, U.S. Forces, European Theater, *Organization and Equipment of Field Artillery Units,* 4.

32. Eisenhower, *Bitter Woods,* 423.

33. "347th Infantry," in *Historical and Pictorial Record of the 87th Infantry Division,* 63.

34. Headquarters, 87th Infantry Division, General Order No. 10, RG 407, NARA, 1.

35. Headquarters, 87th Infantry Division, After Action Report, Jan. 1945, RG 407, NARA, 4 (photocopy in CARL).

36. Headquarters, 345th Infantry Regiment, "History of the 345th," Jan. 1945, RG 407, NARA, 8.

37. Headquarters, 87th Infantry Division, After Action Report, Jan. 1945, 6.

38. Headquarters, 345th Infantry Regiment, "History of the 345th," Jan. 1945, 10.

39. "History of the 346th Infantry," in *Historical and Pictorial Record of the 87th Infantry Division,* 65.

40. Reeves, "Artillery in the Ardennes," 139.

41. Cavanaugh, *Krinkelt-Rocherath,* 135.

42. 2nd Infantry Division, *Combat History,* 95.

43. While the flying conditions necessary for high-altitude bombing and those required for aerial artillery observation may not be exactly the same, a report prepared after the war states: "the first limitation on the use of air power in the European Theater is that on 25 percent of all days it was not possible for the Eighth Air Force to complete bombardment operations. In winter, as many as 10 to 15 days a month were non-operational." U.S. Strategic Bombing Survey, *Weather Factors in Combat Bombardment Operations*, 20.

44. Cole, *Ardennes*, 674.

45. U.S. Adjutant General's Office, *Army Battle Casualties and Nonbattle Deaths in World War II*, 92. Despite the higher-than-normal rate of casualties among forward observers in the Ardennes Campaign, infantrymen in the European theater incurred casualties at an even faster rate. The Adjutant General's Office report indicates that during the entire war, the ratio of U.S. infantrymen to artillerymen who died in battle was about thirteen to one, while in the European theater alone, the ratio was about seventeen to one. Ibid., 48–54. Peter Mansoor notes this disparity in casualties between the two combat arms: "one of the reasons American artillery units were so effective was that they took very few casualties (except for their forward observers)." *GI Offensive in Europe*, 251.

46. Headquarters, 87th Infantry Division, After Action Report, Jan. 1945, 1.

47. Cole, *Ardennes*, 145–46.

48. Ibid., 670–71.

49. Trask, *AEF and Coalition Warmaking*, 19.

Chapter 9

1. Headquarters, 345th Infantry Regiment, "History of the 345th," Feb. 1945, RG 407, NARA, 1–2.

2. McGhee, *Golden Acorn Memories*, 43–44.

3. James R. McGhee, videotaped interview by John R. Walker, Mount Vernon, Ill., May 10, 1999, in author's possession.

4. Bricker to Walker, Sept. 30, 1995.

5. McGhee, *Golden Acorn Memories*, 45.

6. Headquarters, 334th Field Artillery Battalion, "History," sec. 8, "Losses in Action," RG 407, NARA, 1.

7. Headquarters, 87th Infantry Division, S-2 Record of Enemy Activity, Feb. 6, 1945, RG 407, NARA.

8. On February 11, Workman was severely wounded while trying to maintain the only line of communication between the company command post and an infantry platoon holding the high ground. He died a short time

later in a hospital in Belgium. For his heroic efforts, the army awarded him the Bronze Star posthumously. Headquarters, 87th Infantry Division, General Order No. 94, RG 407, NARA, 1; "Dies of Wounds," *Canton (Ohio) Repository,* Mar. 21, 1945, 25.

9. McGhee, *Golden Acorn Memories,* 47–49.

10. Donald L. Walker, conversation with author, November 23, 1988.

11. Headquarters, 87th Infantry Division, General Order No. 219, RG 407, NARA, 1; General Order No. 247, ibid., 1.

12. 345th regimental history file, Fort Jackson and Europe, 1941–50, Loren L. Morrison Papers, U.S. Army Military History Institute, Carlisle, Pa., 52.

13. Gus Epple to John R. Walker, Sept. 22, 1999, in author's possession.

14. 345th regimental history file, Fort Jackson and Europe, 1941–50, Morrison Papers, 52–53.

15. Epple to Walker, Sept. 22, 1999.

16. 345th regimental history file, Fort Jackson and Europe, 1941–50, Morrison Papers, 53.

17. Donald L. Walker to John R. Walker, Aug. 7, 1967.

18. Headquarters, 345th Infantry, "History of the 345th," Feb. 1945, 5–6.

19. Pancoast, "Foot Soldier," 9.

20. Lane W. Barton, Jr., e-mail message to author, Feb. 23, 2000, in author's possession.

21. Headquarters, 334th Field Artillery Battalion, Unit History, Feb. 28, 1945, RG 407, NARA, 1.

22. Headquarters, 345th Infantry, "History of the 345th," Feb. 1945, 6; Headquarters, 334th Field Artillery Battalion, Unit History, Feb. 28, 1945, 1.

23. Headquarters, 334th Field Artillery Battalion, Unit History, Feb. 28, 1945, 1–2.

24. Headquarters, 345th Infantry, "History of the 345th," Feb. 1945, 8.

25. 345th regimental history file, Fort Jackson and Europe, 1941–50, Morrison Papers, 58–61.

26. General Metz notes that not only could the Americans monitor their radio transmissions and decode them via Ultra, but the radios the Germans were using in the Schnee Eifel also did not perform with a high degree of dependability. He adds: "The German portable radio set often did not function because of their technical insufficiency for just such (heavily forested and mountainous) terrain." Richard Metz, *Artillery in the Defensive Battles in the Eifel, along the middle Rhine, in the Wester Wald, in the Siegerland, and dur-*

ing the Destruction of the Ruhr Pocket in the Sector of the Fifth Panzer Army Headquarters and Fifteenth Army Headquarters, from January until the Middle of April, 1945, trans. Charles E. Weber (N.p., Oct. 8, 1948), 42–43, copy at MSTL.

27. "History of the 346th Infantry," in *Historical and Pictorial Record of the 87th Infantry Division,* 65–66.

28. Raymond R. Jemc, "Forty Years Ago," Jemc Papers, n.d., U.S. Army Military History Institute, Carlisle Barracks, Pa., 30–31.

29. Headquarters, 87th Infantry Division, General Order No. 123, RG 407, NARA, 1.

30. House, *Toward Combined Arms Warfare,* 129.

31. "History of the 334th Field Artillery," in *Historical and Pictorial Record of the 87th Infantry Division,* 76–77.

32. McGhee, *Golden Acorn Memories,* 56–57.

33. James R. McGhee, e-mail message to Wayne Gehrt, Aug. 7, 2000, copy in author's possession.

34. Wayne Gehrt to 87th Division Association website, Aug. 5, 2000, copy in author's possession.

35. According to the morning reports, Gazvoda was wounded on February 27. Battery A, 334th Field Artillery Battalion, 87th Infantry Division, Morning Reports, Mar. 1945, NPRC.

36. Headquarters, 87th Infantry Division, General Order No. 244, RG 407, NARA, 1–2.

37. Donagh O'Hara, telephone conversation with author, Nov. 11, 2006.

38. Headquarters, 87th Division Artillery, S-2 Record of Enemy Activity, RG 407, NARA, Mar. 1, 1945.

39. Headquarters, 336th Field Artillery Battalion, "History of the 336th Field Artillery Battalion, Narrative: Month of March 1945," RG 407, NARA, 15.

40. "History of the 346th Infantry," in *Historical and Pictorial Record of the 87th Infantry Division,* 67.

41. 345th regimental history file, Fort Jackson and Europe, 1941–50, Morrison Papers, 78, 81.

42. Jackson, "How the Kaiser Got Unhorsed," 40.

43. 345th regimental history file, Fort Jackson and Europe, 1941–50, Morrison Papers, 84–85.

44. Headquarters, 87th Infantry Division, After Action Report, Mar. 1945, RG 407, NARA, 9.

45. Ibid.

46. "History of the 347th Infantry," in *Historical and Pictorial Record of the 87th Infantry Division,* 87.

47. "Stalwart and Strong," in ibid., 41.

48. Ibid., 43.

49. Headquarters, 87th Infantry Division, After Action Report, Mar. 1945, 13.

50. Headquarters, 87th Infantry Division, General Order No. 262, RG 407, NARA, 2. A *panzerfaust* was the German equivalent of the bazooka, a hand-held rocket launcher used as an antitank weapon.

51. Headquarters, 87th Infantry Division, General Order No. 219, RG 407, NARA, 1.

52. Headquarters, 87th Infantry Division, General Order No. 119, 1.

53. On one occasion, given the option of refusing an upcoming forward-observer mission with Lieutenant Connolly, his men all declared indignantly, "Where 'Fire Mission John' goes, we go, also." Headquarters, 336th Field Artillery Battalion, "History of the 336th Field Artillery Battalion, 135.

54. U.S. Army, *Tank Busters*, 44.

55. Headquarters, 549th AAA Automatic Weapons Battalion, S-2 Periodic Report to S-2 of 87th Division Artillery, Apr. 7–8, 1945, RG 407, NARA.

56. 345th regimental history file, Fort Jackson and Europe, 1941–50, Morrison Papers, 95–97.

57. Donald E. Welever, videotaped interview by John R. Walker, Wadsworth, Ohio, Feb. 9, 2000, in author's possession.

58. 345th regimental history file, Fort Jackson and Europe, 1941–50, Morrison Papers, 97. Although this incident has no known connection to forward observers, at Tambach members of the 345th Infantry allegedly shot sixteen enemy prisoners in cold blood because the young German soldiers had reportedly feigned surrender, then opened fire as the Americans approached them. In his novel *Private*, a fictional account of his experiences with the 345th Infantry, author Lester Atwell describes the incident in detail. Various veterans of the 345th have also written to corroborate details of the incident in more-recent editions of the 87th Division Association's quarterly newsletter, *Golden Acorn News*.

59. Headquarters, 334th Field Artillery Battalion, Unit History, Apr. 1945, RG 407, NARA, 3.

60. Headquarters, 334th Field Artillery Battalion, "History," sec. 8, "Losses in Action."

61. Headquarters, 334th Field Artillery Battalion, Unit History, Apr. 1945, 5.

62. Ibid.

63. Headquarters, 334th Field Artillery Battalion, S-3 Annex to Unit Report No. 5, Apr. 1945, RG 407, NARA, 3.

64. Headquarters, 334th Field Artillery Battalion, S-3 Annex to Unit Report No. 6, May 1945, ibid., 10.

65. Headquarters, 334th Field Artillery Battalion, Unit History, May 1945, ibid., 1.

Chapter 10

1. Headquarters, 37th Infantry Division, General Order No. 46, RG 407, NARA, 4.

2. H. Gordon, "Letter to the Editor," 68.

3. Kahn, "Something Rotten in the Fruit Salad," 21.

4. Ibid.

5. Bernard E. Trainor, "Ribbons over the Heart: Valor or a Bullet's Target?" *Akron (Ohio) Beacon Journal,* May 21, 1996, A11.

6. Editors of the Boston Publishing Society, *Above and Beyond,* 2.

7. Senate Committee on Veterans' Affairs, *Medal of Honor Recipients,* 1079.

8. Mikaelian, *Medal of Honor,* xxv.

9. Taggart, *Third Infantry Division in World War II,* 319.

10. Senate Committee on Veterans' Affairs, *Medal of Honor Recipients,* 651.

11. 63rd Infantry Division Association, *Chronicles,* 16–33.

12. Ibid., 160–61.

13. Ibid., 168.

14. Senate Committee on Veterans' Affairs, *Medal of Honor Recipients,* 665–66.

15. Paul W. Vermillion to John R. Walker, Feb. 24, 2004. Apparently, doctors at the field hospital performed a tracheotomy on Robinson, but he inadvertently pulled the tube out of his trachea and suffocated.

16. Ibid.

17. 63rd Infantry Division Association, *Chronicles,* 169–70.

18. Jacobs, *Heroes of the Army,* 156–57.

19. *Second United States Infantry Division in Korea,* 53.

20. Jacobs, *Heroes of the Army,* 166.

21. *Second United States Infantry Division in Korea,* 54–55.

22. Jacobs, *Heroes of the Army,* 166.

23. Senate Committee on Veterans' Affairs, *Medal of Honor Recipients,* 748–49.

24. Maraniss, *They Marched into Sunlight,* 210–11.

25. Shelton, *Beast Was Out There*, 246.

26. *First Infantry Division in Vietnam*, 101.

27. Maraniss, *They Marched into Sunlight*, 210.

28. Ibid., 210–11.

29. Senate Committee on Veterans' Affairs, *Medal of Honor Recipients*, 832–33.

30. Maraniss, *They Marched into Sunlight*, 274.

31. Senate Committee on Veterans' Affairs, *Medal of Honor Recipients*, 833.

32. Maraniss, *They Marched into Sunlight*, 265.

33. Ibid., 274.

34. Senate Committee on Veterans' Affairs, *Medal of Honor Recipients*, 833.

35. Ibid.

36. Shelton, *Beast Was Out There*, 169.

37. Maraniss, *They Marched into Sunlight*, 275.

38. Shelton, *Beast Was Out There*, 169.

Chapter 11

1. Gen. William DePuy later remarked that "his most important duty as a battalion commander was to get his field artillery forward observers to the next hilltop from which they could direct fires onto the Germans." McGrath, *Fire for Effect*, 66.

2. Weigley, *Eisenhower's Lieutenants*, 28.

3. Overy, *Why the Allies Won*, 226.

4. Cooling, *Development of Close Air Support*, 304, 310, 333. Jonathan House describes air support of ground operations during World War II as "Air-Ground (Non)Cooperation." *Toward Combined Arms Warfare*, 130.

5. Bradley and Blair, *General's Life*, 183.

6. Cooling, *Development of Close Air Support*, 264, 282.

7. Scales, *Firepower in Limited War*, 12–13.

8. McGrath, *Fire for Effect*, 75.

9. Edward G. Gibbons, Jr., "Why Johnny Can't Dismount: The Decline of America's Mechanized Infantry Force," 1995, School of Advanced Studies Monograph, CARL, PDF, http://cgsc.cdmhost.com/cdm/ref/collection/p4013coll3/id/1050, p. 7. Allan Millett and Williamson Murray observe, "although artillery-infantry and artillery-armor coordination received high marks from friend and foe alike, tank and infantry cooperation in infantry divisions showed chronic defects in the European theater of operations." *Military Effectiveness*, 69.

10. House, *Toward Combined Arms Warfare,* 129.

11. Peter Mansoor argues: "In World War II, the most important aspect of [American] tactical maneuver often was that it brought friendly units into a position where they could use fire support assets to destroy the enemy." *GI Offensive in Europe,* 3. Despite his brief tenure as a commissioned officer in the U.S. Army in 1945, my father had some understanding and appreciation of the significance of combined-arms operations. After the January 1968 Tet Offensive in Vietnam, and having seen televised reports of the attack on the U.S. embassy in Saigon, he wrote in a letter to me during my last month in Vietnam: "The news in the papers is all very distressing. Not the fact that the VC should launch infiltration attacks alone, but that they can be so damn successful without air support, artillery support, armor or even motor transport for supplies and munitions." Donald L. Walker to John R. Walker, Feb. 2, 1968, in author's possession.

12. Scott McMeen, an officer in the U.S. Army, identifies field artillery's primary mission during World War I as direct support, which "included all fires delivered in proximity to and support of front line infantry." McMeen writes that this had changed little by the time of World War II, when artillery commanders then defined their primary mission as a "direct support mission" to provide "immediate fire support for infantry regiments and brigades." "Field Artillery Doctrine Development," 15–16.

13. Jonathan B. A. Bailey, "The First World War and the Birth of Modern Warfare," in Knox and Murray, *Dynamics of Military Revolution,* 135.

14. Scott McMeen argues that, with few exceptions, U.S. Army field-artillery doctrine used during World War II differed little from that of World War I. He attributes the "quantum leap" to portable field radios and new fire-direction techniques. "The artillery school did not invent new procedures for ground observers, but simply adapted procedures developed in World War I for aerial observation and correction of fire to ground observation." This argument is sound but fails to attribute improvements in the ability to provide spontaneous fire support to using a system that constantly brought new targets under observation. McMeen admits the importance of forward observers to the success of artillery during World War II, writing, "prewar doctrine failed to recognize the greatly increased demand for ground observers that emerged during World War II." This he attributes primarily to the fact that infantry required more observers to sufficiently cover the wider frontages and secondly to the increased demands for fire requests resulting from much improved communications technology and fire-direction procedures. Again, this is a solid rationale but ignores the fact that the new system was using mobile rather than stationary observers. "Field Artillery Doctrine Development," 37, 59.

15. Gisske to Walker, Oct. 9, 2001. In addition, once German artillery crews spotted Allied observation planes, they quickly ceased firing to avoid revealing their positions. They knew that if they did not, the aerial observer would quickly put them out of action. Cummings, *Grasshopper Pilot*, 48.

16. John McGrath notes: "[A]lthough combat was equally intense, the terrain in the Pacific war was far different from that of Europe. The weather was more extreme in the tropics and thick jungle vegetation made targeting more difficult and fratricide easier." *Fire for Effect*, 77.

17. The study concludes: "Because of the weather, 25 per cent of all days were non-operational for the Eighth and 37 per cent of all days were non-operational for the Fifteenth Air Force. In addition, due to weaknesses in weather forecasting and necessary conservatism in dangerous weather conditions, only about 95 per cent of operational days were exploited." U.S. Strategic Bombing Survey, *Weather Factors in Combat Bombardment Operations*, 2, 20.

18. Gisske to Walker, Oct. 9, 2001. The situation for forward observers serving in the Pacific theater was basically the same. On New Georgia they went through the entire campaign without relief because they were routinely assigned to new battalions as the old ones were relieved. Headquarters, Division Artillery, "Narrative Report of the 37th Division Artillery, Battle of New Georgia," Aug. 11, 1943, APO 37, RG 407, NARA, 3.

19. McManus, *Deadly Brotherhood*, 193.

20. "Heroes All," 523.

21. Ibid., 523.

22. Carver to Walker, Mar. 13, 2000, in author's possession.

23. Bernard Trainor to John R. Walker, June 1, 1996, in author's possession.

24. Robert Schroeder to John R. Walker, Sept. 21, 1999, in author's possession.

25. Although Patton was an old cavalryman and an early advocate of armor, in 1943 he told Sen. Henry Cabot Lodge, "I believe in heavy field artillery." Rickard, *Patton at Bay*, 23.

26. Yamashita Tatsup, 6th Field Artillery: "The regiment brought eighteen guns to the sector but nearly all have been knocked out by U.S. artillery fire. The strength of the 1st Battery was about 150 men but this was reduced to twenty by artillery fire and after last night's barrage (24 March) not a single man was left." 37th Infantry Division, Bougainville Artillery Narrative, 15.

27. Editors, "Call It Infantry—the Artillery's View," 57.

Epilogue

1. Field Artillery School, Instruction Memorandum, sec. 1, "The Infantry Regiment and Supporting Artillery (Combat Team) in the Approach March and Attack," Aug. 1943, MSTL, 28.

2. Anderberg and Wolbarsht, *Laser Weapons*, 119–50.

3. Prehar, *Brave Cannons*, 70–71.

4. Karen Steede-Terry, *Integrating GIS and the Global Positioning*, 5.

5. Headquarters, Department of the Army, *Observed Fire*, FM 6–30 (Washington, D.C.: Government Printing Office, July 16, 1991), 3–4.

6. Manz, Weiss, and Landmesser, "PFED, LWTFDS, and GDU-R," 28–29.

7. Wheaton, "Methods for Improving Jungle Radio Communications," 328.

8. Ralston, "2006: State of Field Artillery," 2.

9. White, "Fires Brigade," 15.

10. Stump, "4th ID Stands Up Army's First Fires Brigade," 27.

11. White, "Fires Brigade," 14.

12. Scales, "Transforming the Force," 8.

13. Stricklin, "Field Artillery," 4.

14. Cheek, "Why Can't Joe Get the Lead Out?," 33.

15. That there was urgency in their voice is undeniable. My mother told me that more than once in the first few months after my father returned home from World War II, he awakened her in the night with his loud and excited talking in his sleep, as if he was speaking into a phone, apparently dreaming or perhaps having a nightmare that he was again directing a fire mission.

16. Cheek, "Why Can't Joe Get the Lead Out?," 34–35.

17. Headquarters, Department of the Army, *Fire Support in Combined Arms Operations*. FM 6–20 (Washington, D.C.: Government Printing Office, Sept. 30, 1977), ii.

18. Combat Action Badge, Army.mil, accessed Sept. 16, 2011, http://www.army.mil/symbols/CombatBadges/action.html.

19. Cheryl Boujnida, "First Soldiers Receive Combat Action Badge," June 29, 2005, Military Medals Guide, accessed Sept. 16, 2011, http://military medalsguide.com/First-Soldiers-receive-Combat-Action-Badge.html.

20. Capt. Effrem S. Gibson, deputy public-affairs officer, 1st Infantry Division, Fort Riley, Kansas, indicates that of the three brigades, the majority of awards, sixty-one, went to soldiers serving with the 3rd Brigade at Fort Knox, Kentucky. Gibson, e-mail message to author, Aug. 25, 2011, in author's possession.

21. "Heroes All," 523.

Glossary

abatis. A form of barricade using trees sharpened and laid with their points outward.

aiming point. An easily distinguishable feature or landmark used to calculate an angle of reference to a gun in the required horizontal direction. The essential feature of the aiming point is that it is at a sufficient distance from the gun using it.

Atabrine. A drug used during World War II to prevent malaria.

automatic weapon. Any hand-held or smaller portable weapon capable of firing continuously as long as the trigger is pulled or depressed and with the ammunition.

azimuth. The horizontal angle measured clockwise between a reference direction and the line to an observed or designated point. True azimuth is the angle measured horizontally from the north in clockwise fashion to an observed or designated point. Figuratively, an azimuth is a horizontal angle or direction from a point on a compass.

base point. A point in the target area whose location is known on the ground, on the firing chart, or both. If its location on the ground is known, it must be readily identifiable and should be in the approximate center of the target area, both horizontally and vertically. The registered known point that became the "base point" established the base deflection of zero mils for each firing battery. Before going to the front lines, the forward observer must know the location of the base point.

battalion. During World War II, a typical U.S. Army infantry battalion had three rifle companies, one heavy weapons company, and a head-quarters company; approximately 1,000 men at full strength. There were three battalions to an infantry regiment, and three infantry regi-ments per infantry division. A typical field-artillery battalion included three batteries of four 105-mm howitzers, one battery of twelve 155-mm howitzers, a headquarters battery, and a service battery. Typically com-manded by a lieutenant colonel or colonel.

battery. The smallest tactical unit of field guns. The 1943 TO&E for a field-artillery battalion—motorized, 105-mm, and truck drawn—al-located four howitzers for each of three firing batteries, each battery typically commanded by a captain.

battery commander. The artillery officer, usually a captain, in com-mand of an individual firing battery, usually of four guns. From the early part of the twentieth century until the start of World War II, the com-mander took a position somewhere between the batteries and the target area, which would enable him to observe a target while communicating with his guns. From that point he performed the multiple tasks of deter-mining targets for his battery, computing the firing data, and adjusting the fire. The position selected often presented a tradeoff between the ability to observe and to conduct fire. A commander who was closer to the target than to the battery could observe better, whereas one closer to the battery had an easier time conducting fire.

bolt-action rifle. A weapon that requires the shooter to eject spent shell casings and inject new ones manually by sliding a handle with a knob back and up and then down and forward every time the weapon is fired. Operating the bolt causes the shooter to lose the alignment of the gun's sight, which means he must reacquire the target after every shot.

bracket. The distance between two strikes or series of strikes, one of which is over the target and the other short of it, or one of which is to the right of a target and another to the left. Standard procedure is to send one artillery round on line with the target over, and the next round short. The target is now "bracketed," and subsequent rounds split the bracket until they land within fifty yards of the adjusting point. The forward observer then fires for effect.

breech-loading. A firearm or field piece that is loaded near the rear of the bore or above the trigger mechanism as opposed to one that is loaded at the top of the barrel, or muzzle.

brigade. A military unit typically composed of two to five regiments or battalions. Usually, a brigade is a subunit of a division and is commanded by a brigadier general.

Bronze Star. The U.S. armed forces fourth-highest award to recognize bravery, heroism, or meritorious service. During World War II, the words used to describe a Bronze Star awarded for bravery or heroism typically included "heroic action." Many were also awarded for meritorious service, though these typically lacked a narrative indicating what the soldier had done to deserve the medal. But some veterans were awarded Bronze Stars for meritorious service with narrative descriptions, indicating the awards had been made to recognize deeds of bravery or heroism in combat.

cannon. A field gun firing a projectile that follows a nearly flat trajectory.

cannoneer. A soldier who performs any of the duties connected with operating a field gun and handling its ammunition.

casualty. A member of the armed forces who is lost to active service, especially as a result of being killed, wounded, or captured in combat. A person who becomes a casualty, then, can be either temporarily or permanently lost to duty.

combat arm. Any combatant branch of the military forces; those branches trained to routinely engage an enemy in combat. During World War II, the combat arms of the U.S. Army included infantry, artillery, armor, air, cavalry, and combat engineers.

combined-arms tactics. The tactical use of two or more combat arms working together.

commissioned officer. An officer in the armed forces holding rank by commission from the proper authority. In the U.S. Army, these ranks range from second lieutenant to general of the army.

company. A U.S. Army rifle company during World War II had three rifle platoons and a heavy weapons platoon. At full strength a rifle company totaled just under 200 officers and enlisted men. Typically commanded by a captain.

control. The detailed and usually local direction of the movement, maneuvers, or fire necessary to accomplish missions or tasks assigned.

corps. A tactical subdivision of an army comprising two or more divisions plus auxiliary service troops.

counterbattery fire. An artillery barrage directed against enemy artillery.

defilade. The protection from hostile observation or fire afforded by an obstacle such as a hill, a ridge, a natural depression in the ground, a bank of earth, or even a building.

deflection. The horizontal shift, left or right, in the direction of the barrel of a field piece necessary to place its fire on line with the target.

deviation. The horizontal angle, measured by an observer, between a burst and the target.

direct fire. Artillery fire in which the cannoneers see and take aim at their intended targets.

elevation. The vertical shift of the barrel of a field piece up or down, necessary to enable its fire to reach the exact distance or range from the gun to the target.

enfilading fire. Fire that strikes the flank of a target, usually from a defilade position. To rake a line lengthwise from the side by rifle or shell fire. The targets then typically have little or no cover or protection.

field artillery. Movable artillery capable of accompanying an army into battle.

fire-direction center (FDC). Location where artillerists compute firing data for the guns of a battery. The process is done by determining the precise target location based on the forward observer's location, then computing range and direction to the target from the guns' location. Fire direction is the tactical command of one or more artillery units for the purpose of bringing their fire to bear upon the proper targets at the proper time with the maximum density and surprise.

global-positioning system (GPS). Developed by the Department of Defense, it uses a network of between twenty-four and thirty-two orbiting satellites to transmit precise radio-wave signals enabling receivers to determine their current location, time, and velocity. Although GPS has significant applications for both military and civilian use, it allows soldiers to find objectives in the dark or unfamiliar territory. GPS coordinates can be used to locate the position of an object on the face of the earth to an accuracy of within two yards. The system also allows accurate targeting of various military weapons, including those used by field artillery.

high-angle fire. Fire in which the quadrant elevation exceeds that for maximum range for the charge. This elevation is approximately 800 mils, or 45 degrees or more.

howitzer. A short-barreled field gun with low muzzle velocity firing shells in a relatively high trajectory.

inclination. The extent or degree of incline from a horizontal or vertical position.

indirect fire. Artillery fire directed at an unseen target.

infantry division. The standard infantry division of the U.S. Army in World War II had a minimum of artillery and auxiliary elements organically assigned. It comprised just under 15,000 officers and enlisted men. In addition to three infantry regiments, it generally included three field-artillery battalions of 105-mm howitzers, one 155-mm howitzer battalion, air section, engineer battalion, medical battalion, cavalry reconnaissance troop, military police platoon, quartermaster company, ordnance company, signal company, headquarters company, and division band. Typically commanded by a major general.

liaison officer. An artillery officer stationed at the supported infantry's command post during battle who relays requests for fire support to his artillery battalion. Liaison included communications, cooperation, and coordination with different elements of command.

logistics. The branch of military science having to do with moving, supplying, and quartering troops.

massed fire(s). Fire delivered by multiple batteries on a target simultaneously.

mil. 1/6400 of a circle. Artillery calculations use this basis almost exclusively for determining the angle of deflection to a target.

mortar. A short-barreled cannon, typically light enough to be carried in sections, which throws shells in a very high trajectory.

noncommissioned officer (NCO). An enlisted person in the armed forces appointed by proper authority to any rank above that of private first class and below that of a commissioned officer or warrant officer.

observation post (OP). A position from which one can observe the location and actions of the enemy, usually in proximity to the enemy. A more permanent observation post may be constructed like a bunker to withstand bombardment, or it may simply be a temporary place for optimal observation, such as a hilltop or a church steeple.

ordnance. All weapons and munitions used in warfare and any equipment or supplies used in servicing weapons.

organic. In military organizational terms, a subunit that is assigned as a regular part of a larger unit. For example, during World War I, the composition of U.S. infantry divisions did not include their own aviation units. But during World War II, aviation units were assigned as a permanent (not attached) part of the division organization.

pillbox. An enclosed, fortified gun emplacement, typically made of concrete and steel.

platoon. During World War II, a U.S. Army infantry platoon typically included three squads of thirteen riflemen each and was commanded by a lieutenant.

private first class (PFC). Two steps up from a buck private entering military service and just below the rank of corporal.

quadrant elevation. The vertical angle from horizontal to line of elevation.

range. The distance from a field gun to its target.

regiment. During World War II, a standard U.S. Army infantry regiment had three battalions and included just over 3,000 officers and enlisted men. Typically commanded by a colonel.

registration. Fire delivered to obtain corrections for increasing the accuracy of subsequent rounds.

rifling. The system of spiral grooves inside a gun barrel that impart spin to a projectile, increasing its ability to maintain a straight trajectory.

rolling barrage. An artillery fire delivered on successive lines, advancing immediately ahead of the attacking troops according to a prearranged schedule. Rolling barrages may be employed to support an attack when the locations of hostile dispositions are obscure, to crater the ground, and to orient and guide attacking troops.

set-piece attack. The attackers know the general location and strength of the defenders and the locus of battle is not likely to change once the assault is underway. Set-piece operations are those that generally go according to plan, though in modern combat, few things ever go exactly as planned.

shrapnel. A term used generically to mean the fragments from any exploding shell. Originally, an artillery shell filled with an explosive charge and many small, metal balls, set to explode in the air over a target.

Silver Star. The armed forces' third-highest award to recognize gallantry in action against an armed enemy of the United States.

small arms. Firearms that can be carried and used by a single soldier, such as rifles, carbines, pistols.

sniper. A sharpshooter concealed to harass the enemy by shooting individuals, usually at a long range with a telescopic rifle. In both world wars, the German army referred to its snipers as sharpshooters.

squad. The smallest tactical unit of a rifle company, typically numbering thirteen enlisted men and led by a sergeant.

table of organization & equipment (TO&E). A standard table established for each type of military unit prescribing the number of its officers and enlisted men, the grade and job of each, the proportion of various military occupational specialties, the arrangement of command and staff, and the numbers of each item of equipment to be supplied to this unit. The publication of the first tables of organization and equipment in August 1943 replaced the tables of organization and tables of basic allowance previously used.

tactical doctrine. The various instructions the U.S. Army teaches all members of combat arms for engaging in combat.

target acquisition. The means of locating and bringing fire to bear on a target using indirect fire. The conversion to indirect fire meant that artillery had to have separate elements to find and observe targets.

target-location error. The distance from the burst of a shell to the target by which the forward observer has initially misjudged the location of the target.

time-on-target (TOT). Artillery fire from multiple sources that is synchronized to fall on a particular target simultaneously.

triangular division. An infantry division premised on the association of all its elements, from squad to regiment, in threes. Ideally it presupposed that one element would fix the enemy while a second maneuvered against him, the third acting as a reserve.

Wehrmacht. The armed forces of Nazi Germany, from *wehr* meaning "armed" and *macht* meaning "might." Some writers have used the term interchangeably with reference to the German army, though the German word for army is *Heer.*

Bibliography

Archives and Manuscripts

Battery C Mothers' Club. Alliance, Ohio. "Monthly Compilation of Letters Received from Sons Serving in the 135th Field Artillery Battalion." April 1945. Copy in author's possession.

C.C.O.R. Bulletin. "Lessons Learnt [*sic*] during the Bougainville Campaign." Reference CCOR/55. Combined Arms Research Library, Fort Leavenworth Kans.

Close Support Study Group. "Executive Study to Final Draft." September 12, 1975. Morris Swett Technical Library. U.S. Army Field Artillery Training Center, Fort Sill, Okla.

Field Artillery School. Instruction Memorandum. Sec. 1, "The Infantry Regiment and Supporting Artillery (Combat Team) in the Approach March and Attack." August 1943. Morris Swett Technical Library, Fort Sill, Okla.

Foreman, Charles R., comp. "The Bulge Battle of the Ardennes. First Battalion 345th." Marquette, Mich., n.d. Unpublished battalion history, 345th Infantry Regiment, 87th Infantry Division. Copy in author's possession.

Headquarters, 37th Infantry Division. Report After Action. "Operations of the 37th Infantry Division, Luzon, P.I., 1 November 1944–30 June 1945." (M-1 Operation) Part VI: "Artillery." Combined Arms Research Library. U.S. Army Command and General Staff College, Fort Leavenworth, Kans.

Headquarters, Americal Division. "Intelligence Annex to Operation Report, Americal Division, Bougainville Operation, 28 December 1943 to 29 April 1944." Combined Arms Research Library, Fort Leavenworth, Kans.

"History of the Forward Observer." Morris Swett Technical Library. U.S. Army Field Artillery Training Center, Fort Sill, Okla.

Jemc, Raymond R. "Forty Years Ago." Raymond R. Jemc Papers. U.S. Army Heritage and Education Center. U.S. Army Military History Institute, Carlisle, Pa.

Jones, Sam. "To Hell and Almost Back." [1990]. Sam Jones Papers. U.S. Army Heritage and Education Center. U.S. Army Military History Institute, Carlisle, Pa.

Metz, Richard. "Artillery in the Defensive Battles in the Eifel, along the Middle Rhine, in the Wester Wald, in the Siegfried Line, and during the Destruction of the Ruhr Pocket in the Sector of the Fifth Panzer Army Headquarters and Fifteenth Army Headquarters from January until the Middle of April 1945." Translated by Charles Weber. October 8, 1948. Morris Swett Technical Library. U.S. Army Field Artillery Training Center, Fort Sill, Okla.

Morning Reports. Battery A, 334th Field Artillery Battalion, 87th Division, December 1944. Military Personnel Records. National Personnel Records Center, St. Louis.

Morning Reports. Company L, 345th Infantry Regiment, 87th Division, December 1944. Military Personnel Records. National Personnel Records Center, St. Louis.

Morrison, Loren L. Papers. Folder, 345th Regiment History. Box, Fort Jackson and Europe, 1941–50. U.S. Army Heritage and Education Center. U.S. Army Military History Institute, Carlisle, Pa.

Office of the Chief of Artillery. Records of the Chiefs of Arms, 1878–1943. Record Group 177. National Archives and Records Administration, College Park, Md.

Records of Headquarters, Army Ground Forces, 1916–54. Record Group 337. National Archives and Records Administration, College Park, Md.

U.S. Army Artillery and Missile School. "History of the Field Artillery School." Vol. 2, "World War II." [1944?]. Morris Swett Technical Library, Fort Sill: Okla.

U.S. Army Ground Forces, Field Artillery School. "In the Pacific Areas." Artillery in Combat, 1944–1945, no. 4. Combined Arms

Research Library. U.S. Army Command and General Staff College, Fort Leavenworth, Kans.

U.S. Field Artillery School. "Observer's Check List, Southwest Pacific Area, 9/26–12/23." 1942. Morris Swett Technical Library. U.S. Army Field Artillery Training Center, Fort Sill, Okla.

War Department. Annual Report for 1919. "Report of the Hero Board: Proceedings of the Board of Officers Convened by the Following Order: General Headquarters, American Expeditionary Forces, Office of the Chief of Artillery, 22 March 1919." Morris Swett Technical Library. U.S. Army Field Artillery Training Center, Fort Sill, Okla.

World War II Operations Reports. Records of the Adjutant General's Office, 1917–45. Record Group 407. National Archives and Records Administration, College Park, Md.

Government Documents

Department of the Army. *History of the Combat Medical Badge.* Army Regulation 600-8-22. February 22, 1995.

Field Artillery School. *Field Artillery Military Fundamentals.* Fort Sill, Okla.: Printing Plant, Field Artillery School, 1942.

———. *Guide for the Liaison Officer and Forward Observer.* Fort Sill, Okla.: Field Artillery School, 1970.

———. *Handbook for the Field Artillery Forward Observer.* Fort Sill, Okla.: Field Artillery School, 1970.

———. *Lateral Observation.* Department of Gunnery, Document 12. Fort Sill, Okla.: Field Artillery School Press, 1922.

General Board, U.S. Forces, European Theater. *Condemnation and Replacement of Artillery Tubes in Combat.* Bad Nauheim, Germany, 1945.

———. *Study of the Organization and Equipment of Field Artillery Units.* Study No. 59. Bad Nauheim, Germany, 1945.

Headquarters, Department of the Army. *Direct Support and General Support Maintenance Manual for Range Finder, Fire Control.* TM 9-1240-369-34. Washington, D.C., February 1974.

———. *Fire Support in Combined Arms Operations.* Field Manual 6–20. Washington, D.C., September 1977.

———. *Personnel Selection and Classification: Commissioned Officer Specialty Classification System.* Army Regulation 611-101. Washington, D.C., April 1962.

———. *Principles of Artillery Weapons.* TM 9–3305. Washington, D.C., May 1981.

———. *Table of Organization and Equipment No. 6-155, Field Artillery Battalion 105-mm, Towed, Infantry Division.* Washington, D.C., March 1966.

———. *Tactics, Techniques, and Procedures for Observed Fire.* Field Manual 6–30. Washington, D.C., September 1991.

Militia Bureau. *Questions for National Guard Officers (Field Artillery).* Washington, D.C.: Government Printing Office, 1920.

U.S. Adjutant General's Office. *Army Battle Casualties and Nonbattle Deaths in World War II: Final Report 7 December 1941–31 December 1946.* Prepared by the Statistical and Accounting Branch, Office of the Adjutant General under Direction of Program Review and Analysis Division, Office of the Comptroller of the Army. Washington, D.C.: Department of the Army, 1953.

U.S. Army. *Artillery Ammunition Expenditures as Related to Infantry Casualties.* Commanding General, XII Corps Artillery. Germany, 1945.

———. *Table of Organization and Equipment No. 6–327, Field Artillery Battery Motorized, 105-mm Howitzer, Tractor-Drawn.* Washington, D.C., October 1944.

U.S. Army Command and General Staff College. "Selected Readings in Military History: Evolution of Combined Arms Warfare." Fort Leavenworth, Kans.: Combat Studies Institute, 1983.

U.S. Coast Guard, Treasury Department. *Electronic Navigational Aids: Loran, Radiobeacon, and Radarbeacon Systems and Loran Radio-Direction-Finder and Radar Ship Equipment.* Washington, D.C., 1949.

U.S. Congress. Senate. Committee on Veterans' Affairs. *Medal of Honor Recipients, 1863–1978.* 96th Cong., 1st sess., 1979. Committee Print 3.

U.S. Strategic Bombing Survey. *Weather Factors in Combat Bombardment Operations in the European Theater.* Washington, D.C.: Military Analysis Division, January 1947.

U.S. War Department. *Applied Tactics, Japanese Army: Translation of a Japanese Manual.* Rev. ed., 1938. Washington, D.C.: Pacific Unit M.I.D., October 1943.

———. *Circular No. 269.* Washington, D.C., October 27, 1943.

———. *Field Artillery: Individual and Unit Training Standards.* TM 6–605. Washington, D.C.: Government Printing Office, February 1945.

———. *Field Artillery Field Manual: Tactics and Techniques.* FM 6–20. Washington, D.C.: Government Printing Office, 1940.

————. *Field Artillery Forward Observation.* FM 6–135. Washington, D.C.: Government Printing Office, August 10, 1944.

————. *Field Artillery Gunnery.* FM 6–40. Washington, D.C.: Government Printing Office, June 1945.

————. *Field Artillery Tactics and Techniques, Battalion and Battery, Motorized.* FM 6–101. Washington, D.C.: Government Printing Office, 1944.

————. *Field Artillery Trainer, M3.* Washington, D.C.: Government Printing Office, 1944.

————. *Field Service Regulations: Operations.* FM 100–5. Washington, D.C.: Government Printing Office, 1941.

————. *Infantry Drill Regulations.* FM 22–5. Washington, D.C.: Government Printing Office, August 1941.

————. *The Radio Operator.* TM 11–454. Washington, D.C., 1942.

————. *Tactics and Techniques of Division Artillery and Higher Artillery Echelons.* FM 6–100. Washington, D.C.: Government Printing Office, 1944.

U.S. War Department, General Staff. *Observed Fires.* Washington, D.C.: Prepared under the Direction of the Chief of Field Artillery, 1941.

Books

2nd Infantry Division. *Combat History of the Second Infantry Division.* Nashville, Tenn.: Battery Press, 1979

37th Infantry Division Pictorial History. Camp Polk, La., 1937.

63rd Infantry Division Association. *The 63rd Infantry Division Chronicles, June 1943 to September 1945.* N.p., 1996.

Addams, Michael C. C. *The Best War Ever: America and World War II.* Baltimore: Johns Hopkins University Press, 1994.

Adkin, Mark. *The Charge: Why the Light Brigade Was Lost.* London: Leo Cooper, 1996.

Altobello, Brian. *Into the Shadows Furious: The Brutal Battle for New Georgia.* Novato, Calif.: Presidio Press, 2000.

Aluit, Alfonso. *By Sword and Fire: The Destruction of Manila in World War II 3 February–3 March 1945.* Manila: National Commission for Culture and the Arts, 1994.

Anderberg, Bengt, and Myron L. Wolbarsht. *Laser Weapons: The Dawn of a New Military Age.* New York: Plenum, 1992.

Appy, Christian G. *Working Class War: American Combat Soldiers and Vietnam.* Chapel Hill: University of North Carolina Press, 1993.

Astor, Gerald A. *A Blood-Dimmed Tide: The Battle of the Bulge by the Men Who Fought It.* New York: Dell, 1994.

———. *Crisis in the Pacific: The Battles for the Philippine Islands by the Men Who Fought Them.* New York: Donald I. Fine Books, 1996.

Atwell, Lester. *Private.* Landsdowne, Pa.: A and A Publishing, 1997.

Babcock, Leslie Edwards. *Elements of Field Artillery.* Princeton: Princeton University Press, 1925.

Bailey, J. B. A. *Field Artillery and Firepower.* Oxford: Military Press, 1989.

———. *Field Artillery and Firepower.* Annapolis: Naval Institute Press, 2004.

Balkowski, Joseph. *Beyond the Beachhead: The 29th Infantry Division in Normandy.* Harrisburg, Pa.: Stackpole, 1989.

Barger, Charles J. *Radio Equipment of the Third Reich, 1933–1945.* Boulder, Colo.: Paladin, 1991.

Barnett, Correlli, ed. *Old Battles and New Defences: Can We Learn from Military History?* London: Brassey's Defence Publishers, 1986.

Beebe, Gilbert W., and Michael E. De Bakey. *Battle Casualties: Incidence, Mortality, and Logistic Considerations.* Springfield, Ill.: Charles C. Thomas, 1952.

Beightler, Robert S. *Major General Robert S. Beightler's Report of the Activities of the 37th Infantry Division, 1940–1945.* N.p., 1946.

Bellamy, Chris. *Red God of War: Soviet Artillery and Rocket Forces.* Washington, D.C.: Brassey's Defence Publishers, 1986.

Birkhimer, William E. *Organization, Administration, Materiel, and Tactics of the Artillery, United States Army.* 1884. Reprint, New York: Greenwood, 1968.

Bradley, Omar N., and Clay Blair. *A General's Life: An Autobiography by the General of the Army.* New York: Simon and Schuster, 1983.

Braim, Paul F. *The Test of Battle: The American Expeditionary Forces in the Meuse-Argonne Campaign.* Newark, N.J.: University of Delaware Press, 1987.

Brookesmith, Peter. *Sniper: Training, Techniques, and Weapons.* New York: St. Martin's, 2006.

Buchan, John. *The Battle of the Somme.* New York: Grosset and Dunlap, 1917.

Caldwell, William H. *Field Artillery Officer's Notes.* New York: E. P. Dutton, 1918.

Carter, Russell Gordon. *The 101st Field Artillery AEF, 1917–1919.* Boston: Houghton Mifflin, 1940.

Cavanaugh, William C. C. *Krinkelt-Rocherath: The Battle for the Twin Villages.* Norwell, Mass.: Christopher Publishing, 1986.

Chang, Iris. *The Rape of Nanking: The Forgotten Holocaust of World War II.* New York: Penguin, 1998.

Clodfelter, Michael. *Warfare and Armed Conflicts: A Statistical Reference to Casualty and Other Figures, 1618–1991.* Vol. 2., *1900–1991.* Jefferson, N.C.: McFarland, 1992.

Close, Jacob. *Memoirs of a Groundpounder.* N.p., c1999.

Cogdan, Don, ed. *Combat WWII European Theater of Operations.* New York: Arbor House, 1987.

Colby, John. *War from the Ground Up: The 90th Division in World War II.* Austin: Nortex, 1991.

Cole, Hugh M. *The Ardennes: The Battle of the Bulge.* United States Army in World War II: The European Theater of Operations. Washington, D.C.: Center of Military History, U.S. Army, 1994.

———. *The Lorraine Campaign.* United States Army in World War II: The European Theater of Operations. Washington D.C.: Historical Division, Department of the Army, 1950.

Combat History of the 44th Infantry Division, 1944–1945. Atlanta: Albert Love Enterprises, 1945.

Connaughton, Richard, John Pimlott, and Duncan Anderson. *The Battle for Manila.* London: Bloomsbury, 1995.

Cooke, James J. *Pershing and His Generals: Command and Staff in the AEF.* Westport, Conn.: Praeger, 1997.

Cooling, Benjamin Franklin, ed. *Case Studies in the Development of Close Air Support.* Washington, D.C.: Office of Air Force History, U.S. Air Force, 1990.

Cooper, John P., Jr. *The History of the 110th Field Artillery with Sketches of Related Units.* Baltimore: War Records Division, Maryland Historical Society, 1953.

Craven, Wesley Frank, and James Lea Cate. *The Army Air Force in World War II.* Vol. 5, *The Pacific: Matterhorn to Nagasaki.* Chicago: University of Chicago Press, 1953.

The Cross of Lorraine: A Combat History of the 79th Infantry Division, June 1942–December 1945. Baton Rouge: Army-Navy Publishing, 1946.

Cummings, Julian William, with Gwendolyn Kay Cummings. *Grasshopper Pilot: A Memoir.* Kent, Ohio: Kent State University Press, 2005.

Dastrup, Boyd L. *The Field Artillery History and Sourcebook.* Westport, Conn.: Greenwood, 1994.

———. *King of Battle: A Branch History of the U.S. Army's Field Artillery.* Fort Monroe, Va.: Office of the Command Historian, U.S. Army Training and Doctrine Command, 1992.

————. *Modernizing the King of Battle, 1973–1991*. U.S. Army Field
Artillery Center and School Monograph Series. Fort Sill, Okla.:
Office of the Command Historian, U.S. Army Field Artillery
Center and School, 1994. Reprint, Washington, D.C.: Center of
Military History, U.S. Army, 2003.

Dillon, Lester R., Jr. *American Artillery in the War with Mexico, 1846–
1848*. Reprint, Austin, Tex.: Presidential, 1975.

Doubler, Michael D. *Closing with the Enemy: How GIs Fought the War in
Europe, 1944–1945*. Lawrence: University Press of Kansas, 1994.

Editors of the Boston Publishing Society. *Above and Beyond: A History of the Medal of Honor from the Civil War to Vietnam*. Produced
in Cooperation with The Congressional Medal of Honor Society.
Boston: Boston Publishing, 1985.

Ehret, Charles F., and Julia Ehret Granton. *As if on Automatic Pilot:
The Battle of Medelsheim, World War II, while a Soldier Reflects*. Clar-
endon Hills/Hinsdale, Ill.: Clockwatcher Books, 1999.

Eisenhower, John S. D. *The Bitter Woods: The Dramatic Story—Told at
All Echelons—from Supreme Commander to Squad Leader—of the
Crisis that Shook the Western Coalition: Hitler's Surprise Ardennes
Offensive*. New York: G. P. Putnam's Sons, 1969.

————. *So Far from God: The War with Mexico, 1846–1848*. New York:
Random House, 1989.

Elstob, Peter. *Hitler's Last Offensive*. London: Secker and Warburg,
1971.

*The Fifth Infantry Division in the ETO: Iceland, England, Ireland, France,
Germany, Luxembourg, Czechoslovakia, Austria*. Prepared by the
Fifth Division Historical Section, Headquarters, Fifth Infantry
Division. Atlanta: Albert Love, 1945. Reprint, Nashville, Tenn.:
Battery, 1981.

The First Infantry Division in Vietnam, 1965–1970. Paducah, Ky.:
Turner Publishing, 1993.

Ford, William Wallace. *Wagon Soldier*. North Adams, Mass.: Excelsior
Printing, 1980.

Frankel, Stanley A. *The 37th Infantry Division in World War II*. Washington, D.C.: Infantry Journal Press, 1948.

Fraser-Tytler, Neil. *Field Guns in France: With a Howitzer Battery in
the Somme, Arras, Messines, & Passchendaele, 1915–1918*. Brighton,
UK: Tom Donovan Publishing, 1922.

Gailey, Harry A. *Bougainville, 1943–1945: The Forgotten Campaign*.
Lexington: University Press of Kentucky, 1991.

Gajdusek, Robert E. *Resurrection: A War Journey: A Chronicle of the
Events during and following the Attack on Fort Jeanne D'Arc at Metz,*

France, by F Company of the 379th Regiment of the 95th Infantry Division, November 14–21, 1944. Notre Dame, Ind.: University of Notre Dame Press, 1997.

Garrison, Webb. *Friendly Fire in the Civil War: More Than 100 True Stories of Comrade Killing Comrade.* Nashville, Tenn.: Rutledge Hill, 1999.

General Staff. *Tactics of the German Army.* Vol. 1, *Defence and Withdrawal.* Prepared under the direction of the Chief of the Imperial General Staff. [London]: War Office, 1944.

Gilbert, Adrian. *Sniper: The Skills, Weapons, and the Experience.* New York: St. Martin's, 1995.

Green, Constance McLaughlin, Harry C. Thomson, and Peter C. Roots. *The Ordnance Department: Planning Munitions for War.* The United States Army in World War II: The Technical Services. Washington, D.C.: Office of the Chief of Military History, Department of the Army, 1955.

Greenfield, Kent Roberts, Robert R. Palmer, and Bell I. Wiley. *The Organization of Ground Combat Troops.* United States Army in World War II: The Army Ground Forces. Washington, D.C.: Historical Division, Department of the Army, 1947.

Grotelueschen, Mark. *The AEF Way of War: The American Army and Combat in World War I.* New York: Cambridge University Press, 2007.

———. *Doctrine under Trial: American Artillery Employment in World War I.* Westport, Conn.: Greenwood, 2001.

Gudmundsson, Bruce. *On Artillery.* Westport, Conn.: Praeger, 1993.

Hallahan, William H. *Misfire: The History of How America's Small Arms Have Failed Our Military.* New York: Charles Scribner's Sons, 1994.

Hammel Eric. *Munda Trail: The New Georgia Campaign.* New York: Orion Books, 1989.

Hardison, Richard M. *An Artillery Captain's Personal War: Caissons across Europe.* Austin, Tex.: Eakin, 1990.

Hartcup, Guy. *The Effect of Science on the Second World War.* New York: St. Martin's, 2000.

Hazlett, James C., Edwin Olmstead, and M. Hume Parks. *Field Artillery Weapons of the Civil War.* Newark, N.J.: University of Delaware Press, 1988.

Henne, Charles A. *Battle Story: A Narrative History, 1940–1945.* 3 vols. Battle History of the 3rd Battalion, 148th Infantry. N.p., 1988–92.

———. *The New Georgia Campaign: Operation Toenails.* Battle History of the 3rd Battalion, 148th Infantry. N.p., 1991.

Hesketh-Pritchard, H. *Sniping in France: With Notes on the Scientific Training of Scouts, Observers, and Snipers.* London: Leo Copper, 1994.

Hewes, James E., Jr. *From Root to McNamara: Army Organization and Administration, 1900–1963.* Washington, D.C.: Center for Military History, 1975.

An Historical and Pictorial Record of the 87th Infantry Division in World War II, 1942–1945. Reprint, 87th Division Association, 1988.

Historical Annual: National Guard and Naval Militia of the State of Ohio, 1938. Baton Rouge, 1938.

A History of the United States Twelfth Armored Division, 15 September 1942–17 December 1945. Reprint, Nashville, Tenn.: Battery, 1978.

Hogg, Ian V. *German Artillery of World War II.* London: Greenhill Books, 2002.

———. *A History of Artillery.* New York: Ballantine Books, 1974.

Hogg, O. F. G. *Artillery: Its Origin, Heyday, and Decline.* Hamden, Conn.: Archon Books, 1970.

Holmes, Richard. *Acts of War: The Behavior of Men in Battle.* New York: Free Press, 1986.

Holmsten, Richard B. *Ready to Fire: Memoir of an American Artillery-man in the Korean War.* Jefferson, N.C.: McFarland, 2003.

House, Jonathan M. *Combined Arms Warfare in the Twentieth Century.* Lawrence: University Press of Kansas, 2001.

———. *Toward Combined Arms Warfare: A Survey of 20th-Century Tactics, Doctrine, and Organization.* Combat Studies Institute, Research Survey 2. Fort Leavenworth, Kans.: U.S. Army Command and General Staff College, 1984.

Hughes, B. P. *Firepower: Weapons Effectiveness on the Battlefield, 1630–1850.* New York: Charles Scribner's Sons, 1983.

Infantry School. *Infantry in Battle.* Washington, D.C.: Infantry Journal, 1939.

Jacobs, Bruce, *Heroes of the Army: The Medal of Honor and Its Winners* New York: W. W. Norton, 1956.

Johnson, Curt. *Artillery: The Big Guns Go to War.* London: Octopus Books, 1975.

Jones, Archer. *Civil War Command and Strategy: The Process of Victory and Defeat.* New York: Free Press, 1992.

Jones, K. P. *F.O. Forward Observer.* New York: Vantage, 1989.

Kedzior, Richard W. *Evolution and Endurance: The U.S. Army Division in the Twentieth Century.* Santa Monica, Calif.: Rand, 2000.

Kemp, Anthony. *The Unknown Battle: Metz, 1944.* New York: Stein and Day, 1981.

Kennett, Lee. *G.I. The American Soldier in World War II.* New York: Charles Scribner's Sons, 1987.

Kerns, Raymond C. *Above the Thunder: Reminiscences of a Field Artillery Pilot in World War II.* Kent, Ohio: Kent State University Press, 2009.

Kirk, John, and Robert Young, Jr. *Great Weapons of World War II.* New York: Walker, 1990.

Knox, MacGregor, and Williamson Murray, eds. *The Dynamics of Military Revolution, 1300–2050.* New York: Cambridge University Press, 2001.

Langellier, John P. *Redlegs: The U.S. Artillery from the Civil War to the Spanish-American War, 1861–1898.* Mechanicsburg, Pa.: Stackpole, 1998.

Langlois, Hippolyte. *Field Artillery in Cooperation with Other Arms.* N.p., 1892.

Lee, Jay McIlvaine. *The Artillerymen: The Experiences and Impressions of an American Artillery Regiment in the World War, 129th Field Artillery, 1917–1919.* Kansas City, Mo.: Press of Spencer Printing, 1920.

Linderman, Gerald. *The World within War: The American Combat Experience in World War II.* New York: Free Press, 1997.

Love, Terry M. *L-Birds: American Combat Liaison Aircraft of World War II.* New Brighton, Minn.: Flying Books, 2001.

Lyons, Michael, ed. *World War II: A Short History.* Upper Saddle River, N.J.: Pearson, Prentice Hall, 2004.

Major, James Russell. *The Memoirs of an Artillery Forward Observer, 1944–1945.* Manhattan, Kans.: Sunflower University Press, 1999.

Maki, Howard T. *A Nine Years' Journey: World War II and Korea, 1943–1952.* Brimley, Mich.: H. T. Maki, 1998.

Mansoor, Peter R. *The GI Offensive in Europe: The Triumph of American Infantry Divisions, 1941–1945.* Lawrence: University Press of Kansas, 1999.

Manucy, Albert. *Artillery through the Ages: A Short Illustrated History of Cannon, Emphasizing Types Used in America.* 1949. Reprint, Washington, D.C.: National Park Service, 1985.

Maraniss, David. *They Marched into Sunlight: War and Peace Vietnam and America, October 1967.* New York: Simon and Schuster, 2003.

Mauldin, Bill. *Up Front.* Reissued facsimile production, New York: W. W. Norton, 2000.

Maurey, Eugene, Jr. *Forward Observer.* Chicago: Midwest Books, 1994.

McClean, Donald B., ed. *Japanese Artillery: Weapons and Tactics.* Wickenburg, Ariz.: Normount Technical Productions, 1973. Reprint of U.S. War Department, Military Intelligence Division. *Japanese Field Artillery.* Special Series 25. Washington, D.C., 1944.

McGhee, James R. *Golden Acorn Memories: After Half a Century.* Mount Vernon, Ill., 1995.

McGrath, John J. *Fire for Effect: Field Artillery and Close Air Support in the U.S. Army.* Fort Leavenworth, Kans.: Combat Studies Institute, 2010.

McKenney, Janice E. *The Organizational History of Field Artillery, 1775–2003.* Washington, D.C.: Center of Military History, 2007.

McManus, John C. *The Deadly Brotherhood: The American Combat Soldier in World War II.* Novato, Calif.: Presidio, 1998.

McWhiney, Grady, and Perry D. Jamieson. *Attack and Die: Civil War Military Tactics and the Southern Heritage.* Tuscaloosa: University of Alabama Press, 1982.

Menning, Bruce W. *Bayonets before Bullets: The Imperial Russian Army, 1861–1914.* Bloomington: Indiana University Press, 1992.

Mikaelian, Michael, *Medal of Honor: Profiles of America's Military Heroes from the Civil War to the Present* (New York: Hyperion, 2002).

Miller, John, Jr. *Cartwheel: The Reduction of Rabaul.* The United States Army in World War II: The War in the Pacific. Washington, D.C.: Office of the Chief of Military History, Department of the Army, 1959.

———. *Guadalcanal: The First Offensive.* The United States Army in World War II: The War in the Pacific. Washington, D.C.: Historical Division, Department of the Army, 1949.

Millett, Allan, and Williamson Murray, eds. *Military Effectiveness.* Vol. 3, *The Second World War.* Mershon Center Series on Defense and Foreign Policy. Boston: Unwin Hyman, 1988.

———. *Military Innovation in the Interwar Period.* New York: Cambridge University Press, 1996.

———. *A War to be Won: Fighting the Second World War, 1937–1945.* Cambridge, Mass.: Harvard University Press, 2000.

Morita, Hiroaki. *The Nation's Most Decorated Military Unit: The 100th/442nd Regimental Combat Team.* USAWC Military Studies Program Paper. Carlisle Barracks, Pa.: U.S. Army War College, 1992.

Nichols, David, ed. *Ernie's War: The Best of Ernie Pyle's World War II Dispatches.* New York: Random House, 1986.

Norris, John. *Artillery: A History.* Stroud, Gloucestershire: Sutton Publishing, 2000.

Odom, William. *After the Trenches: The Transformation of U.S. Army Doctrine, 1918–1939*. College Station: Texas A&M University Press, 1999.

Ohl, John Kennedy. *Minuteman: The Military Career of General Robert S. Beightler*. Boulder, Colo.: Lynne Rienner, 2001.

Ott, David Ewing. *Field Artillery, 1954–1973*. Vietnam Studies. Washington, D.C.: Department of the Army, 1975.

Overy, Richard. *Why the Allies Won*. London: Jonathan Cape, 1995.

Paret, Peter, ed. *Makers of Modern Strategy from Machiavelli to the Nuclear Age*. With the collaboration of Gordon A. Craig and Felix Gilbert. Princeton, N.J.: Princeton University Press, 1986.

Pay, D. R. *Thunder from Heaven: Story of the 17th Airborne Division, 1943–1945*. Reprint, Nashville, Tenn.: Battery, 1980.

Peek, Chet. *The Taylorcraft Story*. Terre Haute, Ind.: Sunshine House, 1992.

Percin, Alexandre. *Le massacre de notre infanterie, 1914–1918*. Paris: Albin Michael, 1921.

Phillips, Robert H. *To Save Bastogne*. New York: Stein and Day, 1983.

Pictorial History of the 37th Division, United States Army, Camp Shelby, Mississippi, 1940–1941. Atlanta: Army-Navy Publishing, 1941.

Ping, Donald H. *The Smoking War: An Autobiography*. Warren, Ind.: D. H. Ping, 1990.

Prehar, Bohdan. *Brave Cannons: WW II Relic in Vietnam, 1st Battalion, 92nd Artillery*. Austin, Tex., 2003.

Prentiss, Augustin M. *Chemicals in the War: A Treatise on Chemical Warfare*. New York: McGraw-Hill, 1937.

Raines, Edgar F., Jr. *Eyes of the Artillery: Origins of Modern U.S. Army Aviation in World War II*. Washington, D.C.: Center of Military History, 2000.

Remini, Robert V. *The Battle of New Orleans*. New York: Viking, 1999.

Rentz, John N. *Bougainville and the Northern Solomons*. Washington, D.C.: Historical Section, Division of Public Information, Headquarters, Marine Corps, 1948.

Richter, Donald C. *Chemical Soldiers: British Gas Warfare in World War I*. Lawrence: University Press of Kansas, 1992.

Rickard, John Nelson. *Patton at Bay: The Lorraine Campaign, September to December 1944*. Westport, Conn.: Praeger, 1999.

Ripley, Warren. *Artillery and Ammunition of the Civil War*. New York: Promontory, 1999.

Rogers, Horatio. *World War I through My Sights*. Rafael, Calif.: Presidio, 1976.

Rouquerol, Gabriel. *The Tactical Employment of Quick-Firing Artillery.* Translated by F. DeB Radcliffe. 1903.

Rusiecki, Stephen M. *The Key to the Bulge: The Battle for Losheimergraben.* Westport, Conn.: Praeger, 1996.

Scales, Robert H., Jr. *Firepower in Limited War.* Washington, D.C.: National Defense University Press, 1990.

Seaton, Albert. *The German Army, 1933–1945.* New York: St. Martin's, 1982.

The Second United States Infantry Division in Korea, 1950–1951. Tokyo: Toppan Printing, 1952.

Senich, Peter R. *The German Sniper, 1914–1945.* Boulder, Colo.: Paladin, 1982.

Sexton, William Thaddeus. *Soldiers in the Sun: An Adventure in Imperialism.* Washington, D.C.: 1944. Reprint, Freeport, N.Y.: Books for the Libraries Press, 1971.

Shaw, Henry I., Jr., and Douglas T. Kane. *Isolation of Rabaul.* Vol. 2 of *History of the U.S. Marine Corps Operations in World War II.* Washington, D.C.: Historical Branch, G-3 Division, Headquarters, U.S. Marine Corps, 1963.

Shelton, James E., *The Beast Was Out There: The 28th Infantry Black Lions and the Battle of Ong Thanh, Vietnam, October, 1967.* Chicago: Cantigny First Division Foundation, 2002.

Shrader, Charles R. *Amicicide: The Problem of Friendly Fire in Modern War.* Fort Leavenworth, Kans.: U.S. Army Command and General Staff College, 1985.

Simpson, Andy. *Hot Blood and Cold Steel: Life and Death in the Trenches in the First World War.* London: Tom Donovan Publishing, 1993.

Sligh, Robert Bruce. *The National Guard and National Defense: The Mobilization of the Guard in World War II.* Westport, Conn.: Praeger, 1992.

Smith, Robert Ross. *The Approach to the Philippines.* The United States Army in World War II: The War in the Pacific. Washington, D.C.: Office of the Chief of Military History, Department of the Army, 1953.

———. *Triumph in the Philippines.* The United States Army in World War II: The War in the Pacific. Washington, D.C.: Center of Military History, 1993.

Snow, William J. *Signposts of Experience: World War Memoirs.* Washington, D.C.: United States Field Artillery Association, 1941.

Spiller, Roger J., ed. *Combined Arms in Battle since 1939.* Fort Leavenworth, Kans.: U.S. Army Command and General Staff College Press, 1992.

Staff of Field Artillery Journal. *Field Artillery Guide.* Washington, D.C.: U.S. Field Artillery Association, 1942.

Stannard, Richard. *Infantry: An Oral History of an American Infantry Battalion.* New York: Twayne Publishers, 1993.

Stanton, Shelby L. *Order of Battle, U.S. Army, World War II.* Novato, Calif.: Presidio, 1984.

Sunderland, Riley. *Outline History of the FAS.* Fort Sill, Okla.: Field Artillery School, 1944.

Steede-Terry, Karen. *Integrating GIS and the Global Positioning System.* Redlands, Calif.: ESRI, 2000.

Supreme Commander of Allied Powers. *Reports of General MacArthur.* Charles A. Willoughby, editor in chief. Vol. 1, *The Campaigns of MacArthur in the Pacific.* 1966. Reprint, Washington, D.C.: Center of Military History, 1994.

Taggart, Donald G., ed. *History of the Third Infantry Division in World War II.* Washington, D.C.: Infantry Journal, 1947.

Thompson, Paul W., ed. *How the Jap Army Fights.* New York: Penguin, 1942.

Toland, John. *Battle: The Story of the Bulge.* New York: Random House, 1959.

Trask, David F. *The AEF and Coalition Warmaking, 1917–1918.* Lawrence: University Press of Kansas, 1993.

U.S. Army. *Tank Busters: The History of the 607th Tank Destroyer Battalion in Combat on the Western Front.* Munich, Ger.: Knorr and Hirth, 1945.

U.S. War Department. *Handbook on Japanese Military Forces.* TM-E30-480. Washington, D.C.: Government Printing Office, 1944. Reprint, with a new introduction by David Isby and afterword by Jeffrey Ethell. Baton Rouge: Louisiana State University Press, 1991.

Van Creveld, Martin. *Fighting Power: German and U.S. Army Performance, 1939–1945.* Westport, Conn.: Greenwood, 1982.

Wacker, Albercht. *Sniper on the Eastern Front: The Memoirs of Sepp Allerberger, Knights Cross.* Barnsley, South Yorkshire: Pen and Sword Military, 2005.

Weigley, Russell F. *The American Way of War: A History of United States Military Strategy and Policy.* New York: Macmillan, 1973.

———. *Eisenhower's Lieutenants: The Campaign of France and Germany, 1944–1945.* Bloomington: Indiana University Press, 1981.

———. *A History of the United States Army.* New York: Macmillan, 1967.

Whiting, Charles. *Massacre at Malmedy: The Story of Joachim Peiper's Battle Group, Ardennes, December 1944.* New York: Stein and Day, 1971.

———. *Siegfried: The Nazis' Last Stand.* New York: Stein and Day, 1982.

Williams, Mary. H., comp. *Chronology, 1941–1945.* United States Army in World War II, Special Studies. Washington, D.C.: Office of the Chief of Military History, Department of the Army, 1960.

Wilson, John B. *Maneuver and Firepower: The Evolution of Divisions and Separate Brigades.* Washington, D.C.: Center of Military History, U.S. Army, 1998.

Woods, David L. *A History of Tactical Communications.* New York: Arno, 1974.

Worley, Marvin L. *A Digest of New Developments in Army Weapons, Tactics, Organization, and Equipment.* Harrisburg, Pa.: Military Service Publishing, 1958.

Wright, B. C. *The 1st Cavalry Division in World War II.* Tokyo: Tappan Printing, 1947.

Zaloga, Steve. *U.S. Field Artillery of World War II.* New York: Osprey, 2007.

Articles

Adkins, Walter D. "This New Fire Direction Technique." *Field Artillery Journal* (December 1941).

Anderson, John B. "Are We Justified in Discarding Pre-War Methods of Training?" *Field Artillery Journal* (April–June 1919).

"The Annual Report of the Chief of Field Artillery for 1933." *Field Artillery Journal* (January–February 1934).

Barker, Harold R. "Artillery Operations of the New Georgia Campaign." *Field Artillery Journal* (August 1944).

Berman, Charles, trans. "Japanese Operations in China, Digested from Articles in the Krasnaya Zveyzda, April 23 and 24, 1938, and February 8, 1939." *Field Artillery Journal* (November 1939).

Berry, Lucien G. "Some Difficulties in Supporting an Infantry Division." *Field Artillery Journal* (January–February 1934).

Bird, John F. "The Forward Observer." *Field Artillery Journal* (July 1941).

Blanchard, C. C. "Control of the Fire of a Battalion by a Single Forward Observer." *Field Artillery Journal* (July–August 1933).

Boyle, Conrad A. "Has the Close-Support Problem Been Solved?" *Field Artillery Journal* (September–October 1939).

Burns, John J. "The Light Artillery Battalion Functioning as a Fire Unit." *Field Artillery Journal* (July–August 1933).

Bush, Richard D. "Forward Observation in Africa." *Field Artillery Journal* (October 1943).

Carlson, Raymond. "Jap POWs' View of Our Artillery." *Field Artillery Journal* (April 1946).

Case, R. H. "Artillery Support in Attack." *Field Artillery Journal* 42 (May–June 1935).

Casey, John F., Jr. "An Artillery Forward Observer on Guadalcanal." *Field Artillery Journal* (August 1943).

"Centennial of Ohio's 135th Field Artillery." *Field Artillery Journal* (May–June 1939).

Cheek, Gary H. "Why Can't Joe Get the Lead Out?" *FA Journal* (January–February 2003).

Chief of Field Artillery. "The Annual Report of the Chief of Field Artillery for 1933." *Field Artillery Journal* (February 1933).

Cocklin, Robert F. "Bougainville—1944." *Field Artillery Journal* (July 1944).

Dransfield, Thomas. "Selling Artillery Support to the Infantry." *Field Artillery Journal* (June 1944).

Drummond, N. L. "On the Way." *Infantry Journal* (July 1944).

Editors. "Call It Infantry—the Artillery's View." *Infantry Journal* (November 1946).

———. "A German Reflects upon Artillery: Interrogation of Karl Thoholte, General der Artillerie." *Field Artillery Journal* (December 1945).

———. "Reasons for the Success of German Field Artillery." *Field Artillery Journal* (December 1941).

Ferris, John W. "Jungle Communications." *Field Artillery Journal* (March 1944).

Foreman, Chuck. "The Other Side of the Hill." *Golden Acorn News* 38, no 3 (October 1996).

Franke, G. H. "Liaison with and Fire Support of the Front Line Infantry." *Field Artillery Journal* (September–October 1925).

From a Letter to Maj. Gen. Robert M. Danford, U.S.A., ret. "The Team on New Georgia by an Infantry Battalion Commander." *Field Artillery Journal* (November 1943).

Fuller, Ralph M. "Liaison in the Jungle." *Field Artillery Journal* (September 1944).

"German Conduct of Fire." *Field Artillery Journal* (September 1943).

Gibbons, Ulrich G. "Let's Use Forward Observation." *Field Artillery Journal* (May 1946).

Gildart, Robert C. "Artillery on New Georgia." *Field Artillery Journal* (February 1944).

Gordon, Harold J., Jr. "Letter to the Editor of Infantry Journal." *Infantry Journal* (November 1947).

Gordon, John. "Battle in the Streets—Manila, 1945." *Field Artillery* 4 (August 1990).

Guenther, John C. "Artillery in the Bougainville Campaign." *Field Artillery Journal* 35 (June 1945).

Haines, Howard F. "Division Artillery in the Battle of New Georgia." *Field Artillery Journal* (November 1943).

Hays, George P. "Fire Direction of Artillery Supporting Infantry." *Field Artillery Journal* (May 1921).

Hearn, J. Richard. "Early Luzon Experience." *Field Artillery Journal* 35 (June 1945).

"Heroes All: High Tribute to Artillery Observers—Taken from the History of the 96th Infantry Division, Soon to be Published by the Infantry Journal." *Field Artillery Journal* 36 (September 1946).

Hoehle, Donald F. "I'll Always Be Grateful to Them." *Golden Acorn News* 40, no. 2 (June 1998).

Jackson, Phil H. "How the Kaiser Got Unhorsed." *Golden Acorn News* 43, no. 1 (March 2001).

Jones, Lloyd E. "Infantry-Artillery Liaison in Combat." *Field Artillery Journal* (September–October 1930).

Jones, Spencer. "The Influence of Horse Supply upon Field Artillery in the American Civil War." *Journal of Military History* 74 (April 2010).

Kahn, E. J. Jr. "Something Rotten in the Fruit Salad." *Infantry Journal* (May 1946).

Kirkland, Robert O. "Orlando Ward and the Gunnery Department: The Development of the F.D.C." *Field Artillery Journal* (June 1995).

Lanza, Conrad H. "The Artillery Support of the Infantry in the AEF." *Field Artillery Journal* (January–February 1936).

Loring, David, Jr. "Instruction in Field Artillery Tactics for Officers of Infantry." *Field Artillery Journal* (November–December 1925).

Manz, Paul C., Jeffrey L. Weiss, and John A. Landmesser. "PFED, LWTFDS, and GDU- R: You Want Tactical Handhelds? We've Got Tactical Handhelds." *FA Journal* (May–June 2003).

Maurey, Eugene, Jr. "Sound Locations." *Field Artillery Journal* (February 1945).

McKenney, Janice. "More Bang for the Buck in the Interwar Army: The 105-mm Howitzer." *Military Affairs* (April 1978).

Merten, Fred W., trans. "Japan Modernizes Her Field Artillery." *Field Artillery Journal* (September 1937).

Milotta, David E. "Who Won the Medals?" *Field Artillery Journal* (November 1947).

Morgan, Prentice G. "The Forward Observer." *Military Affairs* 23 (1959).

Neuffer, William. "What Lessons in the Employment of Field Artillery Should Be Deduced from The Experiences of the Russo-Japanese War?" Translated by G. LeR. Irwin. *Field Artillery Journal* (April–June 1911).

Norris, Frank W. "With Mediums." *Field Artillery Journal* (March 1945).

Ostler, Harry R. "In France . . . with 105s." *Field Artillery Journal* (March 1945).

Pancoast, Edwin C. "Foot Soldier, 1943–1946." *Golden Acorn News* 43, no. 2 (June 2001).

Parkhurst, Charles D. "The Artillery at Santiago." *Journal of the United States Artillery* (March–April 1899).

Ralston, David C. "2006: State of Field Artillery." *FA Journal* (November–December 2006).

Ratliff, Frank G. "The Field Artillery Fire Direction Center." *Field Artillery Journal* (May–June 1950).

Reeves, Joseph R. "Artillery in the Ardennes." *Field Artillery Journal* (March 1946).

Revermann, Hans. "Artillery Observation," from *Voelkischer Beobachter*. *Field Artillery Journal* (August 1940).

Ridings, Arthur. "Friendly Fire." *Golden Acorn News* 42, no. 1 (March 2000).

Ritchie, Robert B., "Elementary Artillery for the Doughboy." *Infantry Journal* (November 1943).

Rodgers, Archibald M. "The 136th Field Artillery Battalion." *Field Artillery Journal* (December 1945).

Rogers, Archibald. "The Cagayan Valley Operation." *Field Artillery Journal* 36 (February 1946).

Sackton, Frank J. "Battle Notes of Division Artillery on Luzon." *Field Artillery Journal* 35 (September 1945).

Scales, Robert H., Jr. "Transforming the Force: From Korea to Today." Interview by Patricia Slayden Hollis. *FA Journal* (July–August 2000).

Shem, Carl A. "Notes on a Regimental March, Truck-Drawn." *Field Artillery Journal* (November–December 1935).

Smyth, Eugene R. "Fighting the Nips—With 105s." *Field Artillery Journal* (September 1945).

Snow, William, trans. "Method of Observing Fire by Use of Lateral Observers." *Field Artillery Journal* (April–June 1911).

Stricklin, Toney. "Field Artillery: Relevant, Trained and Ready Two Years Later." *FA Journal* (July–August 2001).

Stump, Michael M. "4th I. D. Stands Up Army's First Fires Brigade," *FA Journal* (January–February 2005).

Terraine, John. "Indirect Fire as a Battle Winner/Loser." In Barnett, *Old Battles and New Defences*, 7–31.

Thielen, Bernard. "Some Notes for Forward Observers." *Field Artillery Journal*, April 1942.

Valcourt, David P., Robert T. Bray, and Tommy A. Williamson. "2004: The State of Field Artillery." *FA Journal* (November–December 2004).

Van Dyning, Dale. "FO Comments from the Beachhead." *Field Artillery Journal* (July 1944).

Van Horne, Richard W. "Short-Range Firing against the Siegfried Line." *Field Artillery Journal* (February 1945).

Von Ondarza, M. "The Forward Observer: A Tactical Problem Illustrating the Actions of a Forward Observer, and How Artillery-Infantry Liaison Operates in the German Army." *Field Artillery Journal* (April 1941).

Wells, Bruce H. "Fifteen Days—The Defense of Damulaan." *Field Artillery Journal* (April 1945).

Wheaton, J. W. "Methods for Improving Jungle Radio Communications." *Field Artillery Journal* (June 1945).

White, Samuel R. "The Fires Brigade: Not Your Daddy's FFA HQ." *FA Journal* (November–December 2005).

Wogan, John B. "Air Observation of Field Artillery." *Field Artillery Journal* (February 1941).

Wood, John S., comp. "Artillery Notes from Foreign Journals." *Field Artillery Journal* (May 1936).

———. "Digest of a Course in Artillery Given in the School of War in France: French Artillery Doctrine." *Field Artillery Journal* 22 (July–August 1922).

———. "The Liaison Problem." *Field Artillery Journal* (July–August 1933).

Wrockloff, G. E. "The Best Radio Wavelength for the Field Artillery." *Field Artillery Journal* (July–August 1933).

Yeuell, Donovan. "Random Reflections on Light Artillery in Combat." *Field Artillery Journal* (February 1945).

Theses, Dissertations, and Other Academic Papers

Barnes, Peter R. "Command, Control, Communications, and Automation Needs for the Combined Arms Team." Master's thesis, U.S. Army Command and General Staff College, 1994.

Brigham, Wesley C. "Practical Means for Improving Liaison between the Infantry and the Field Artillery." Fort Leavenworth, Kans., Command and General Staff School, 1932.

Carter, Donald Alan. "From G.I. to Atomic Soldier: The Development of U.S. Army Tactical Doctrine, 1945–1956." PhD diss., The Ohio State University, 1987.

Corkill, William E. "A Study of the Development of the Tactical Employment of Field Artillery since the Napoleonic Wars to Determine the Fundamental Principles of its Employment." CGSS Student Papers, 1930–36. Fort Leavenworth, Kans., Command and General Staff School, 1932.

Dastrup, Boyd L. "Travails of Peace and War: Field Artillery in the 1930s and Early 1940s." From a paper presented to the 1990 Conference of Army Historians, Washington, D.C.

Garay, Stephen L. "The Breach of Intramuros." Research report for the Armored School, Fort Knox, Ky., May 1, 1948.

Ling, David H. "Combined Arms in the Bradley Infantry Platoon." Master's thesis, U.S. Army Command and General Staff College, 1993.

McCaul, Edward Baldwin. "The Soul of Modern Artillery: The Development of the Mechanical Projectile Fuse during the American Civil War." Master's thesis, The Ohio State University, 2000.

McMeen, Scott R. "Field Artillery Doctrine Development, 1917–1945." Master's thesis, U.S. Army Command and General Staff College, 1991.

Meyer, Vincent. "Evolution of Field Artillery Tactics during and as a Result of the World War." Fort Leavenworth, Kans., U.S. Army Command and General Staff College, 1930.

Nesmith, Vardell Edwards, Jr. "The Quiet Paradigm Change: The Evolution of the Field Artillery Doctrine of the United States Army, 1861–1905." PhD diss., Duke University, 1966.

O'Brien, Charles B. "Toward Army Maneuver Transformation." School of Advanced Military Studies Monographs. Fort Leavenworth, Kans.: U.S. Army Command and General Staff College, 2006.

Pierce, Richard Lee. "A Maximum of Support: The Development of U.S. Army Field Artillery Doctrine in World War I." Master's thesis, The Ohio State University, 1993.

Shugart, David Adams. "On the Way: The U.S. Field Artillery in the Interwar Period." PhD diss., Texas A&M University, 2002.

Snodgrass, James G. "Operational Maneuver—from the American Civil War to the OMG: What Are Its Origins and Will It Work Today?" School of Advanced Military Studies Monographs. Fort Leavenworth, Kans., U.S. Army Command and General Staff College, 1988.

Snyder, Paul E. "Revolution or Evolution? Combined Arms Warfare in the Twenty-First Century." Master's thesis, U.S. Army Command and General Staff College, 1999.

Spaulding, O. L. "The New Field Artillery Materiel—Its Characteristics and Powers." Lecture by O. L. Spaulding, Artillery Corps, September 5–6, 1905. Department of Military Art, Infantry and Cavalry School, Fort Leavenworth, Kans., 1905.

Spencer, Frederic W. "A Slow March to Military Effectiveness: The Motorization of the United States Field Artillery." Master's thesis, The Ohio State University, 1991.

Stebbins, Steven Allen. "Indirect Fire: The Challenge and Response in the U.S. Army, 1907–1917." Master's thesis, University of North Carolina, 1993.

Steckel, Glenn A. "The Role of Field Artillery in the Siege on Intramuros, Manila, P.I." Research report for the Armored School, Fort Knox, Ky., May 7, 1948.

Index